Lecture Notes in Physics

Volume 842

For further volumes:
http://www.springer.com/series/5304

The Lecture Notes in Physics

The series Lecture Notes in Physics (LNP), founded in 1969, reports new developments in physics research and teaching—quickly and informally, but with a high quality and the explicit aim to summarize and communicate current knowledge in an accessible way. Books published in this series are conceived as bridging material between advanced graduate textbooks and the forefront of research and to serve three purposes:

- to be a compact and modern up-to-date source of reference on a well-defined topic
- to serve as an accessible introduction to the field to postgraduate students and nonspecialist researchers from related areas
- to be a source of advanced teaching material for specialized seminars, courses and schools

Both monographs and multi-author volumes will be considered for publication. Edited volumes should, however, consist of a very limited number of contributions only. Proceedings will not be considered for LNP.

Volumes published in LNP are disseminated both in print and in electronic formats, the electronic archive being available at springerlink.com. The series content is indexed, abstracted and referenced by many abstracting and information services, bibliographic networks, subscription agencies, library networks, and consortia.

Proposals should be sent to a member of the Editorial Board, or directly to the managing editor at Springer:

Christian Caron
Springer Heidelberg
Physics Editorial Department I
Tiergartenstrasse 17
69121 Heidelberg/Germany
christian.caron@springer.com

Hans Paetz gen. Schieck

Nuclear Physics
with Polarized Particles

 Springer

Hans Paetz gen. Schieck
Institut für Kernphysik
Universität zu Köln
Zülpicher Straße 77
50937 Cologne
Germany
e-mail: h.schieck@t-online.de

ISSN 0075-8450
ISBN 978-3-642-24225-0
DOI 10.1007/978-3-642-24226-7
Springer Heidelberg Dordrecht London New York

e-ISSN 1616-6361
e-ISBN 978-3-642-24226-7

Library of Congress Control Number: 2011938361

Cover design: eStudio Calamar, Berlin/Figueres

Printed on acid-free paper

Springer is part of Springer Science+Business Media (www.springer.com)

In memoriam

This book is dedicated to the memory of **Paul Huber** (1910–1971) who—at the University of Basel—pioneered an entire new field of nuclear physics: polarization phenomena, and with it the series of spin-physics (or polarization-phenomena) conferences and workshops. He was a brilliant teacher of physics, wrote a four-volume physics textbook, and worked in many international organizations such as IUPAP. Many of his former Ph.D. students became internationally renowned researchers and university teachers. Left photo reprinted with permission from AIP, New York. Copyright Physics Today 24(6) p. 71 (1971). Right photo by the author.

Preface

Over 30 years a nuclear spin-polarization program has been maintained at the Institut für Kernphysik of the University of Cologne. The successful implementation of such a program requires not only the necessary equipment much of which has to be developed in-house but it rests on specially instructed and trained collaborators. These come about from undergraduate and graduate students who often start to work in experimental groups already as "mini-researchers", develop into diploma or masters and later into Ph.D. students. Besides the standard education in physics, then by specializing in nuclear physics, and finally in specialized lectures and seminars they may become involved in fields like spin physics. In order to give them a knowledge basis in view of a lack of literature different scripts were written to accompany the lecture topics. One of the scripts was on the formal description of spin polarization, another on polarized-ion sources. It seems worthwhile to collect and conserve this knowledge in the form of a printed lecture note.

This lecture note consists of several parts. A large part is devoted to introducing the formal theory, the description of polarization and of nuclear reactions with polarized particles. Another part describes the physical basis of methods and devices necessary to perform experiments with polarized particles and to measure polarization and polarization effects in nuclear reactions. A brief review of modern applications in medicine and fusion-energy research will conclude the lecture note. However, the many contributions of polarization to the widespread field of nuclear physics, especially nuclear reactions, i.e. its results and achievements in that context can only be touched upon within this more methodical survey.

Especially in the more experimental parts of the lecture note it appears impossible to cite all relevant references completely. Therefore, only original references to important developments of the field or selected references to the more recent literature, preferably containing further more complete references, can be cited here. They have been selected in view of their exemplary (not necessarily priority) value or, when discussing devices of polarization physics, the author will show examples with which he is acquainted in order to introduce the principles and more recent developments. Therefore, the examples are mainly taken from low-energy installations such as tandem-Van-de-Graaff laboratories although the emphasis of present research is shifting to medium- and high-energy nuclear physics and the number of low-energy installations is waning. Consequently the description is entirely non-relativistic and focussed on the energy range from astrophysical energies ($\approx$10 keV) to tens of MeV. Also it is restricted to polarization of hadronic particles i.e. the polarization effects of electrons or γ radiation are not treated.

Acknowledgments

It is a pleasure for me to thank all colleagues and especially my students over many years who not only have worked on improving the theoretical understanding of polarization phenomena but also on developing and improving the technical means and relevant devices which were used successfully for nuclear physics, especially in the field of low-energy few-body reactions.

Contents

Part I Formalism: Description of Spin Polarization

1 Introduction 3
References 6

2 Spin States and Spin Polarization 9
2.1 Measurement Process, Pure and Mixed States, Polarization . . . 10
2.2 Expectation Value and Average of Observables
in Measurements 11
References 12

3 Density Operator, Density Matrix 13
3.1 General Properties of ρ 13
3.2 Distinction Between Pure and Mixed States 14
3.2.1 Pure State 14
3.2.2 Mixed State 15
3.3 Other General Properties of ρ 16
3.4 Examples for Density Matrices 17
3.4.1 Spin $S = 1/2$ 17
3.4.2 Spin $S = 1$ 20
3.4.3 Rotation of a Pure $S = 1$ State 22
3.5 Complete Description of Spin Systems 23
3.6 Expansions of the Density Matrix, Spin Tensor Moments 24
3.6.1 Expansion of ρ in a Cartesian Basis
for Spin $S = 1/2$ 25
3.6.2 Spin $S = 1$ 26
3.6.3 Limiting Values of the Polarization Components 29
3.6.4 Expansion Into Spherical Tensors 29
3.6.5 Example for the Construction of a Set of Spherical
Tensors for $S = 1$ 32
3.6.6 Spin Tensor Moments 34

3.6.7   Spherical Tensors, Density Matrix, and Tensor
        Moments for Spin $S = 1/2$ 34
3.6.8   Density Matrix and Tensor Moments
        for Spin $S = 1$ 35
3.6.9   Polarization of Particles with Higher Spin   37
References   38

4   Rotations, Angular Dependence of the Tensor Moments   39
    4.1   Generalities   39
    4.2   The Description of Rotations by Rotation Operators   39
    4.3   Rotation of the Density Matrix and of the Tensor Moments ...   41
    4.4   Practical Realization of Rotations   43
    4.5   Coordinate System   43
    References   44

Part II   Nuclear Reactions

5   Description of Nuclear Reactions of Particles with Spin   47
    5.1   General   47
    5.2   The M Matrix   48
    5.3   Types of Polarization Observables   50
    5.4   Coordinate Systems   52
          5.4.1   Coordinate Systems for Analyzing Powers   52
          5.4.2   Coordinate Systems for Polarization Transfer   53
          5.4.3   Coordinate Systems for Spin Correlations   54
    5.5   Structure of the M Matrix and Number
          of "Necessary" Experiments   57
    5.6   Examples   60
          5.6.1   Systems with Spin Structure $1/2 + 0 \longrightarrow 1/2 + 0$ ...   60
          5.6.2   Systems with the Spin Structure
                  $1/2 + 1/2 \longrightarrow 1/2 + 1/2$ 61
          5.6.3   The Systems with Spin Structure $\frac{1}{2} + \vec{1}$
                  and Three-Nucleon Studies   62
          5.6.4   The Systems with Spin Structures $\vec{1} + \vec{1}$ and $\frac{\vec{1}}{2} + \frac{\vec{1}}{2}$
                  and the Four-Nucleon Systems   62
          5.6.5   Practical Criteria for the Choice of Observables   63
    References   64

6   Partial Wave Expansion   65
    6.1   Neutral Particles   65
    6.2   Charged Particles   68
    6.3   Computer Codes   69
    References   70

7   Charged-Particle Versus Neutron-Induced Reactions   71
    References   72

Part III   Devices

8   Sources and Targets of Polarized H and D Ions   75
    8.1   Physical Basics: General Introduction   75
    8.2   Hyperfine Structure   77
        8.2.1   HFS of the H Atom   77
        8.2.2   HFS in a Magnetic Field (Zeeman Effect)   78
        8.2.3   Zeeman Splitting of the H Atom   81
        8.2.4   Zeeman Splitting of the D Atom   82
        8.2.5   Calculation of Polarization   83
    8.3   Physics and Techniques of the Ground-State Atomic
        Beam Sources ABS   88
        8.3.1   Production of H and D Ground-State
             Atomic Beams   88
        8.3.2   Dissociators, Beam Formation and Accomodation . . .   88
        8.3.3   State-Separation Magnets: Classical and Modern
             Designs   90
        8.3.4   RF Transitions   93
    8.4   Ionizers   102
        8.4.1   Ionizers: Electron-Bombardment and CBS Designs . . .   102
        8.4.2   Sources for Polarized ^{6,7}Li and ^{23}Na Beams   106
        8.4.3   Optically Pumped Polarized Ion Sources (OPPIS) . . .   107
    8.5   Physics of the Lambshift Source LSS   107
        8.5.1   The Lambshift   107
        8.5.2   Level Crossings and Quench Effect   108
        8.5.3   Enhancement of Polarization   109
        8.5.4   Examples of the Polarization Calculation
             for Different Modes of the LSS   109
        8.5.5   Hydrogen   110
        8.5.6   Deuterium   111
        8.5.7   Production and Maximization of the
             Beam Polarization   114
    8.6   Spin Rotation in Beamlines and Precession in a Wien Filter . . .   121
        8.6.1   Spin Rotation in Beamlines   121
        8.6.2   Spin Rotation in a Wien Filter   123
    8.7   Polarized (Gas) Targets and Storage Cells   126
    References   128

9   Polarization by Optical Pumping 131
      9.1   Principles 131
      9.2   Polarization of ^{3}He.............................. 131
             9.2.1   Polarization by Metastability Exchange........... 132
             9.2.2   Spin Exchange............................. 133
      9.3   Ion Sources for Polarized ^{3}He Beams.................. 135
      References 135

Part IV   Methods

10   Production of Polarization by Other Methods 139
      10.1   Polarized Charged-Particle Beams from Nuclear Reactions . . . 139
      10.2   Polarized Neutrons from Nuclear Reactions.............. 139
      10.3   Spin Filtering: Interaction of Low-Energy
             Neutrons with Hyperpolarized ^{3}He.................... 140
      10.4   Spin Filtering for Polarized Antiprotons 141
      References 141

11   Measurement of Polarization Observables.................. 143
      Reference 144

12   Polarimetry 145
      12.1   Absolute Methods.................................. 145
             12.1.1   Time Reversal and Double Scattering . . . . : 145
             12.1.2   Analytical Behavior of the Scattering Amplitudes . . . 146
             12.1.3   Calibration Points Due to a Special Spin Structure . . . 147
             12.1.4   Calibration Due to Special Conditions 149
             12.1.5   Typical Low-Energy Analyzer Reactions......... 150
             12.1.6   Polarimetry in Polarization-Transfer Experiments.... 153
      12.2   Polarimetry of Atomic (and Molecular) Beams 154
             12.2.1   Breit–Rabi Polarimeters...................... 154
             12.2.2   Lambshift Polarimeters LSP 156
      References 157

Part V   Applications

13   Medical Applications 161
      13.1   Hyperpolarized ^{3}He and ^{129}Xe....................... 161
      References 163

14 "Polarized" Fusion 165
    14.1 Five-Nucleon Fusion Reactions 166
    14.2 Four-Nucleon Reactions 167
        14.2.1 Suppression of Unwanted DD Neutrons 168
    14.3 Status of "Polarized" Fusion 170
    References 172

15 Outlook 175
    References 176

Index 179

Part I
Formalism: Description
of Spin Polarization

Chapter 1
Introduction

Spin is an entirely non-classical property of (elementary) particles [1], p. 198. A reference which deals with many aspects of spin (although the—in my mind—seminal Stern–Gerlach experiment (for details see Sect. 8.1) is never even mentioned) is "The Story of Spin" by Tomonaga [2] from which we cite: "It is a mysterious beast, and yet its practical effect prevails over the whole of science. The existence of spin, and the statistics associated with it, is the most subtle and ingenious design of Nature—without it the whole universe would collapse" (from the translator's (T. Oka) preface, p. vii). In nuclear physics, the nuclear, nucleon, and even quark spins enter in many ways. Examples are: hyperfine interaction, spin-orbit interaction, tensor force, spin–spin nucleon–nucleon interaction, the relation between spin and statistics which is not only the basis of the periodic table, but markedly influences the scattering of identical particles (e.g. ^{12}C on ^{12}C vs. ^{13}C on ^{13}C,) etc.

The measurement of spin-polarization observables in reactions of nuclei and particles always proves to be useful or advantageous when effects of single spin substates are to be investigated. The unpolarized differential cross-section encompasses the averaging over the spin states of the particles in the entrance channel and a summation over those of the exit channel. In this averaging process often details of the interaction will be lost. The reasons for this are:

- In contrast to the cross-section polarization observables generally contain interference terms between different amplitudes leading to higher sensitivity to small admixtures.
- Because nuclear forces are spin dependent in various ways polarization observables allow the separation of the different contributions (example: spin-orbit or tensor force).
- In the so-called "complete" experiments (example: nucleon–nucleon interaction) a minimum number of independent observables to be described by different bilinear combinations of transition amplitudes, are necessary.
- In addition, there are physical questions which can be answered only by polarization experiments. *Examples are*: Parity violation in the nucleon–nucleon

H. Paetz gen. Schieck, *Nuclear Physics with Polarized Particles*,
Lecture Notes in Physics 842, DOI: 10.1007/978-3-642-24226-7_1,
© Springer-Verlag Berlin Heidelberg 2012

Table 1.1 Table of the historical development of several series of spin-polarization/spin-physics conferences and workshops

Year	No. in series	Polarization symposia	Ref.	No. in series	Name	High-energy spin physics conferences	Ref.	No. in series	Polarized beams/ targets workshops	Ref.
1960	1	Basel	[13]							
1965	2	Karlsruhe	[14]							
1970	3	Madison	[3]							
1974				1		Argonne	[15]			
1975	4	Zürich	[16]							
1976				2		Argonne	[17]			
1978				3		Argonne	[18]			
1980	5	Santa Fe	[19]	4		Lausanne	[20]			
1981								1	AnnArbor	[21]
1982				5		Brookhaven	[22]			
1983								2	Vancouver	[23]
1984				6		Marseille	[24]			
1985	6	Osaka	[25]							
1986				7	SPIN 86	Protvino	[26]	3	Montana	[27]
1988				8	SPIN 88	Minneapolis	[28]			
1990	7	Paris	[29]	9	SPIN 90	Bonn	[30]	4	Tsukuba	[31]
1992				10	SPIN 92	Nagoya	[32]			
1993								5	Madison	[33]
1994	8	Bloomington	[34]	11		Bloomington	[35]			

The numbering in the different series was not always consistent

interaction, contribution of the spins of constituent quarks to the spin of the nucleon etc.

The measurement of polarization and analyzing power components or—more general—of "generalized analyzing powers" and the preparation of polarized beams and targets require special knowledge and special skills. All measurements can be reduced to intensity measurements. Nevertheless polarization observables are derived from measurements of intensity differences of the particles prepared depending on their spin in different ways (simplest example: the left-right asymmetry produced in a nuclear reaction which has a spin-dependent differential cross-section, e. g. due to an $(\vec{\ell} \cdot \vec{s})$ force). The transformation properties of the tensors into which the observables are usually expanded enter the description in a decisive way. Since we normally deal with statistical ensembles of particles (e.g. in the form of partially polarized beams and targets, not in the form of pure states) the adequate description is by introducing the density operator (density matrix) expanded into representations suitable for the special situation. Two of these have proven most useful:

Table 1.2 Table of the historical development of the series of HE spin-physics conferences and polarized beams, sources, and targets workshops

Year	No. in series	Name	HE spin conf.	Ref.	No. in series	Name	Workshops	Ref.
1995 [a]		SPIN 95	Protvino	[36]	6		Köln	[37]
1996	12	SPIN 96	Amsterdam	[38]				
1997					7	PST97	Urbana	[39]
1998	13	SPIN 98	Protvino	[40]				
1999					8	PST99	Erlangen	[41]
2000	14	SPIN 2000	Osaka	[42]				
2001 [b]		SPIN 2001	Beijing		9	PST01	Nashville	[43]
2002	15	SPIN 2002	Brookhaven	[44]				
2003					10	PST03	Novosibirsk	[45]
2004	16	SPIN 2004	Trieste	[46]				
2005					11	PST05	Tokyo	[47]
2006	17	SPIN 06	Kyoto	[48]				
2007					12	PST07	Brookhaven	[49]
2008	18	SPIN 08	Charlottesville	[50]				
2009					13	PST09	Ferrara	[51]
2010	19	SPIN 2010	Jülich	[52]				
2011					14	PST11	St. Petersburg	
2012	20	SPIN 2012	Dubna					

[a] not numbered [b] 3rd Circum-Pan-Pacific Symp. on HE Spin Physics

- The Cartesian representation in which the observables transform as Cartesian tensors (e.g. under rotations), more adapted to our normal spatial view of the world.
- The spherical representation in which the transformations are identical to those of the spherical harmonics, leading—among others—to simplifications in the description especially of systems with higher spins.

The present script resulted from a lecture course "Nuclear Physics with Polarized Particles". Its purpose was to make the reader acquainted with the formalism unavoidably connected with spin polarization and its background. General treatments of the basics and formalisms have been published, and the following are recommended for additional reading: Refs. [3] and especially [4] therein, [5] and especially [6] therein [7, 8]. Only conventions generally accepted today will be used. Nuclear reactions will be addressed only so far as the knowledge from an introductory course on nuclear reactions has to be expanded. The wealth of often very detailed and also technical material forces one to a narrow down the selection of the necessary subjects. However, literature references for more advanced subjects will be given. The development of polarization physics (or "spin physics" as it is now often called) can be followed in all details in the proceedings of mainly three series of polarization conferences which are listed in Tables 1.1 and 1.2 . The parallel series of Polarization Symposia and High-Energy Spin Physics Conferences merged after the Bloomington conferences 1984.

The other series concerned workshops dealing predominantly with the tools of polarization physics. Here the tremendous technical developments of polarization physics (sources of polarized ion beams, polarized gas targets, polarimeters etc.) have been documented in proceedings of the workshop series in addition to the original articles which appeared mainly in *Nuclear Instruments and Methods in Physics Research* and *Review of Scientific Instruments*. Additionally, there were conferences or workshops on more specialized subjects such as (solid) polarized targets, target materials and techniques (Saclay 1966, Abingdon 1981, Bad Honnef 1984, and Heidelberg 1991), on the physics and applications of polarized ^{3}He ("HELION"), see e. g. [9–11], polarized antiprotons (the earliest at Bodega Bay [12]), electron polarization etc., as well as conferences with more local importance which will not be listed here.

References

1. Landau, L.D., Lifshitz, E.M.: Quantum Mechanics (Non-relativistic Theory), 3rd edn. Pergamon, New York (1977)
2. Tomonaga, S.I.: The Story of Spin. The University of Chicago Press, Chicago (1997)
3. Barschall, H.H., Haeberli, W. (eds.): Proceedings of the 3rd International Symposium on Polarization Phenomena in Nuclear Reactions, Madison 1970. University of Wisconsin Press, Madison (1971)
4. Darden, S.E.: In: [3] p. 39 (1971)
5. Fick, D. (ed.): Proceedings Meeting on Polarization Nuclear Physics, Ebermannstadt 1973. Lecture Notes in Physics 30. Springer, Berlin (1974)
6. Simonius, M.: In: [5] p. 37 (1974)
7. Ohlsen, G.G., McKibben, J.L., Lawrence, G.P., Keaton, P.W. Jr., Armstrong, D.D.: Phys. Rev. Lett. **27**, 599 (1971)
8. Conzett, H.E.: Rep. Progr. Phys. **57**, 1 (1994)
9. Tanaka, M. (ed.): Proceedings of the 7th International RCNP Workshop on Polarized ^{3}He Beams and Gas Targets and Their Applications "HELION97", Kobe 1997. North Holland, Amsterdam (1997)
10. Heil, W.: Proceedings of the International Conference on Polarized ^{3}He("HELION02"), Oppenheim 2002, on CD-ROM (2003)
11. JCNS Workshop on Trends in Production and Applications of Polarized ^{3}He, Ismaning 2010, IOP J. of Physics: Conference Series (2011)
12. Krisch, A.D., Chamberlain, O., Lin, A.T.M. (eds.): Proceedings of the Workshop on Polarized Antiprotons, Bodega Bay 1987. AIP Conf. Proc. 145, New York (1986)
13. Huber, P., Meyer, K.P. (eds.): Proceedings of the International Symposium on Polarization Phenomena of Nucleons, Basel 1960. Helv. Phys. Acta Suppl. VI, Birkhäuser, Basel (1961)
14. Huber P., Schopper H. (eds.): Proceedings of the 2nd International Symposium on Polarization Phenomena of Nucleons, Karlsruhe 1965. Experientia Suppl. 12, Birkhäuser, Basel (1966)
15. Thomas, G.H. et al. (eds.): Proceedings of the Symposium on High Energy Physics with Polarized Beams, Argonne 1974. Atomic Energy Commission (1974)
16. Grüebler, W., König, V. (eds.): Proceedings of the 4th International Symposium on Polarization Phenomena in Nuclear Reactions, Zürich 1975. Birkhäuser, Basel (1976)
17. Marshak, M.L. (ed.): Proceedings of the Conference on High-Energy Physics with Polarized Beams and Targets, Argonne 1976. AIP Conf. Proc. 35, New York (1976)
18. Thomas, G.H. (ed.): Proceedings of the Conference on High-Energy Physics with Polarized Beams and Targets, Argonne 1978, AIP Conf. Proc. 51, New York (1979)

19. Ohlsen, G.G., Brown, R.E., Jarmie, N., McNaughton, W.W., Hale, G.M. (eds.): Proceedings of the 5th International Symposium on Polarization Phenomena in Nuclear Physics, Santa Fe 1980. AIP Conf. Proc. 69, New York (1981)

20. Joseph, C., Soffer, J. (eds.): Proceedings of the Conference on High-Energy Physics with Polarized Beams and Targets, Lausanne 1980. Experientia Suppl. 38. Birkhäuser, Basel (1981)

21. Krisch, A.D., Lin, A.T.M. (eds.): Proceedings of the Workshop on High Intensity Polarized Proton Ion Sources, Ann Arbor 1981. AIP Conf. Proc. 80, New York (1982)

22. Bunce, G.M. (ed.): Proceedings of the 5th High-Energy Spin Physics Symposium, Upton 1982. AIP Conf. Proc. 95, New York (1983)

23. Roy, G., Schmor, P. (eds.): Proceedings of the International Conference Polarized Proton Ion Sources, Vancouver 1983, AIP Conf. Proc. 117, New York (1984)

24. Soffer, J. (ed.): Proceedings of the Conference on High Energy Spin Physics, Marseille 1984. Les Editions de Physique, 46 (1985)

25. Kondo, M., Kobayashi, S., Tanifuji, M., Yamazaki, T., Kubo, K.-I., Onishi, N. (eds.): Proceedings of the 6th International Symposium on Polarization Phenomena in Nucl. Physics, Osaka 1985. Suppl. J. Phys. Soc. Jpn. 55 (1986)

26. Soloviev, L.D. (ed.): Proceedings of the (1986) Spin Symposium, Protvino/Serpukhov. IHEP (1986)

27. Jaccart, S., Mango, S. (eds.): Proceedings of the International Workshop on Polarized Sources and Targets, Montana 1986. Helv. Phys. Acta 59 (1986)

28. Proceedings of the 8th International Symposium on High-Energy Spin Physics, Minneapolis 1988. AIP Conf. Proc. 187, New York (1989)

29. Boudard, A, Terrien, V. (eds.): Proceedings of the 7th International Conference on Polarization Phenomena in Nuclear Physics, Paris 1990. Colloque de Phys. 51, colloque C6, Les Editions de Physique, Paris (1990)

30. Althoff, K.-H., Meyer, W. (eds.): Proceedings of the 9th International Symposium on High Energy Spin Physics SPIN90, Bonn 1990. Springer, Berlin (1991)

31. Mori Y., (ed.): Proceedings of the Intern. Workshop on Polarized Ion Sources and Polarized Gas Jets, Tsukuba 1990. KEK, Tsukuba (1990)

32. Hasegawa, T., Horikawa, N., Masaike, A., Sawada S. (eds.): Proceedings of the 10th International Symposium On High-Energy Spin Physics (SPIN 92—35th Yamada Conference) 1992, Nagoya, "FRONTIERS OF HIGH ENERGY SPIN PHYSICS" Tokyo (1992)

33. Anderson, L.W., Haeberli W. (eds.): Proceedings of the Workshop on Polarized Ion Sources and Polarized Gas Targets, Madison 1993. AIP Conf. Proc. 293, New York (1994)

34. Stephenson, E., Vigdor, S.E. (eds.): Proceedings of the 8th International Symposium on Polarization Phenomena in Nuclear Physics, Bloomington 1994. AIP Conf. Proc. 339, New York (1995)

35. Heller, K.E., Smith, S.L. (eds.): Proceedings of the 11th International Symposium on High Energy Spin Physics, Bloomington 1994. AIP Conf. Proc. 343, New York (1995)

36. Nurushev, S.B. (ed.): Proceedings of the International Workshop on High-Energy Spin Physics (SPIN-95). Protvino (1996)

37. Sydow, L., Paetz gen. Schieck, H. (eds.): Proceedings of the International Workshop on Polarized Beams and Gas Targets, Köln 1995. World Scientific, Singapore (1996)

38. de Jager, C.W., Ketel, T.J., Mulders, P.J., Oberski J.E.J., Oskam-Tamboezer, M. (eds.): Proceedings of the 12th International Symposium on High Energy Spin Physics (SPIN 96), Amsterdam 1996. World Scientific, Singapore (1997)

39. Holt, R., Miller, M. (eds.): Proceedings of the 7th International Workshop on Polarized Gas Targets and Polarized Beams, Urbana 1997. AIP Conf. Proc. 421, New York (1998)

40. Nurushev, S.B. (ed.): Proceedings of the 13th International Symposium on High-Energy Spin Physics, Protvino 1998. World Scientific, Singapore (1999)

41. Gute, A., Lorenz, S., Steffens, E. (eds.): Proceedings of the International Workshop on Polarized Sources and Targets (PST99), Erlangen 1999. Friedrich-Alexander U. Erlangen-Nürnberg (1999)

42. Hatanaka, K., Nakano, T., Imai, K., Ejiri, H. (eds.): Proceedings of the 14th International Spin Physics Symposium, Osaka 2000. AIP Conf. Proc. 570, New York (2001)
43. Derenchuk V.P., and Przewoski, B.v. (eds.): Proceedings of the 9th International Workshop on Polarized Sources and Targets, Nashville 2001. World Scientific Publishing ISBN 981-02-4917-9 (2002)
44. Makdisi, Y.I., Luccio, A.U., McKay, W.W. (eds.): Proceedings of the 15th International Spin Physics Symposium, Upton, N.Y., and Workshop on Polarized Electron Sources and Polarimeters, Danvers, Mass. 2002. AIP Conf. Proc. 675, New York (2003)
45. Terskhov, A.S., Rachek, I.A., Toporkov D.K. (eds.): Proceedings of the 10th International Workshop on Polarized Beams and Targets, Novosibirsk 2003. Nucl. Instrum. Methods Phys. Res. A **536** (2005)
46. Bradamante, F., Bressan, A., Martin, A., Aulenbacher, K.(eds.): Proceedings of the 16th International Spin Physics. Symposium and Workshop on Pol. Electron Sources and Polarimeters, Trieste 2004. World Scientific, Singapore (2005)
47. Uesaka T., Sakai, H., Yoshimi, A., Asahi, K. (eds.): Proceedings of the 11th International Workshop on Polarized Sources and Targets, Tokyo 2005. World Scientific, Singapore (2007)
48. Imai K., Murakami, T., Saito, N., Tanida, K. (eds.): Proceedings of the 17th International Spin Physics Symposium (SPIN 2006), Kyoto 2006. AIP Conf. Proc. 915, New York (2007)
49. Alessi, J., Bunce, G., Kponou, A., Makdisi, Y.I., Zelenski, A. (eds.): Proceedings of the 12th International Workshop on Polarized Sources and Targets, Brookhaven 2007. AIP Conf. Proc. 980, New York (2008)
50. Crabb, D.G. et al. (eds.): Proceedings of the 18th International Symposium on Spin Physics, Charlottesville 2008. AIP Conf. Proc. 1149, New York (2009)
51. Ciullo, G., Contalbrigo, M., Lenisa, P. (eds.): Proceedings of the 13th International Workshop on Polarized Sources, Targets and Polarimetry (PST09), Ferrara 2009. World Scientific, Singapore (2011)
52. Rathmann, F., Ströher, H. (eds.): Proceedings of the International Spin Conference (SPIN2010), Jülich 2010. Published as Open Access by IOP Conference series (2011)

Chapter 2
Spin States and Spin Polarization

Quantum mechanics deals with statistical statements about the result of measurements on an ensemble of states (particles, beams, targets). In other words: by giving an expectation value of operators it provides probabilitites (better: probability amplitudes) for the result of a measurement on an ensemble. Here two limiting cases can be distinguished. One is the case that our knowledge about the system is complete e.g. when all members of an ensemble are in the same spin state. This state will then be characterized completely by a state vector (ket). A special case is the spin state of a *single* particle which is always completely (spin-)polarized.

Example In the classic Stern–Gerlach experiment [1] (see Chap. 8.1) an inhomogeneous magnetic field (the "polarizer") spatially separates silver atoms according to two projections $m_S = +1/2$ and $m_S = -1/2$ of a spin S $= 1/2$ system.[1] By cutting out one of these partial beams by a stopper the remaining ensemble will be completely spin-polarized, see Fig. 8.1. The general state vector of such a system (as well as of each single particle in the beam) is

$$|\Psi\rangle = a|\uparrow\rangle + b|\downarrow\rangle \tag{2.1}$$

with $|\uparrow\rangle$ and $|\downarrow\rangle$ denoting the basis states which characterize such a system completely, i.e. the UP or DOWN spin projections with respect to one direction, e.g. that of the magnetic field. a and b are the probability amplitudes for the two basis states with the normalization $|a|^2 + |b|^2 = 1$. Like for every two-state system the dimension of this system is N $= 2$ ($=2S + 1$) and therefore the description of this system is complete if $|\uparrow\rangle$ and $|\downarrow\rangle$ are orthogonal. The choice of the quantization axis is in principle arbitrary. It is, however, obvious that—though e.g. the occupation of the substates under the rotation transformation of this axis is not changed—the phase relation between $|\uparrow\rangle$ and $|\downarrow\rangle$ and therefore the description will be changed. $|\Psi\rangle$ is a

---

[1] At the time of this completely unexpected discovery neither the notion of a spin S nor of its $2S + 1$ projections was possible. Whenever we discuss spin in general the symbols S and m_S are used. The symbols J and m_J are used for the electron spin of atoms (and sometimes for nuclear spin states), and I, m_I are reserved for the nuclear spins.

H. Paetz gen. Schieck, *Nuclear Physics with Polarized Particles*,
Lecture Notes in Physics 842, DOI: 10.1007/978-3-642-24226-7_2,
© Springer-Verlag Berlin Heidelberg 2012

coherent superposition of the basis states. Denoting as usual the polar and azimuthal angles of the "spin" by β and ϕ

$$\frac{a}{b} = \frac{\cos \frac{\beta}{2}}{e^{i\phi} \sin \frac{\beta}{2}} \tag{2.2}$$

The solution of this equation shows that one has a state in which the spins of all particles point into a well-defined direction (β, ϕ). In the Stern–Gerlach case we can check the direction dependence of a and b with a second Stern–Gerlach magnet (the "analyzer") which is rotatable around the beam axis: the intensities behave as $\sin^2(\beta/2)$ or $\cos^2(\beta/2)$, respectively. This double Stern–Gerlach experiment has been used to introduce a number of basic features of quantum mechanics such as the anti-commutation and commutation relations for the spin and other properties of the spin operators, the projection-operator formalism etc., see e.g. [2, 3, 4, 5].

In classical optics this is quite analogous to the behaviour of light polarization only that the angle $\beta/2$ has to be replaced by β when rotating polarization filters. (The factor 1/2 characterizes the spin as an entirely non-classical phenomenon).

A state for which we have complete knowledge about all particles of the ensemble is a *pure state*.

The other limiting case is that where the magnetic field of the Stern–Gerlach system is turned off. Then for symmetry reasons the spins of all particles of the ensemble will point in all spatial directions with equal probability (completely unpolarized ensemble). It is clear that this system cannot be described solely by one state vector and also that for such a system we do not know in which direction the spin of each individual particle is pointing. It is clearly a state of non-maximal information (even, as we shall see, it is a state of minimal information). A Stern–Gerlach analyzer would not detect any dependence on direction. There is also no fixed phase relation between basis states. As in classical (statistical) physics the adequate description of the state is that of an incoherent weighted superposition of pure states. In contrast to the case of the pure states it should not matter with reference to which direction the pure UP/DOWN basis states have been defined.

2.1 Measurement Process, Pure and Mixed States, Polarization

From the foregoing discussion the recipe of how to describe a general non-pure state (mixed state, mixed ensemble or "mixture") results: one has to state with which probability p_i (not probability amplitude, but statistical weight!) a number of pure states contribute to the mixture. In detail:

- $p_i \geq 0$ and $\sum_i p_i = 1$
- Choose a "basis" of pure states described by state vectors $\Psi^{(i)}$ which do not necessarily have to be orthogonal!

Example: one could combine one pure state completely polarized in $+x$ direction with a contribution of $P_1 = 20\%$ with another one completely polarized in the $-z$ direction with a contribution of $P_2 = 80\%$

- The number of these states needs not be equal to the dimension N of the state space (spin space) but can be larger! E.g. one could imagine a partially polarized ensemble of spin 1/2 particles produced from three pure states in the following way: $P_1 = 30\%$ of the particles fully polarized in the $+z$ direction, $P_2 = 40\%$ fully polarized in the $+x$ and $P_3 = 30\%$ fully polarized in the $-y$ direction.
- In the limit an infinite number of subsystems each completely polarized in arbitrary directions can be imagined. If their statistical weights would all be equal the measured spin polarization would be zero.

Example: Modern Stern–Gerlach magnets as they are used in atomic-beam sources for polarized particle beams are multipole magnets, see Sect. 8.3.3 (quadrupoles and sextupoles). In such magnets all field directions around the z axis appear with equal weight because of rotational symmetry. This means that even though the magnet focuses and therefore selects all atoms in one spin substate and therefore produces partial beams completely polarized with respect to one field direction the polarization in the magnet interior will be zero. This results from the *incoherent* superposition of the *pure, i.e. fully polarized* subsystems with equal weights. Only after guiding the atoms (adiabatically!) into a field region with a dominant field direction will there be a net polarization.

2.2 Expectation Value and Average of Observables in Measurements

Carrying out a number of measurements of an observable **A** on a (generally) mixed ensemble results in an expectation value which is the statistical ensemble average of the quantum-mechanical expectation values $\langle \Psi^{(i)}|A|\Psi^{(i)}\rangle$ with respect to the pure states $|\Psi^{(i)}\rangle$ present (or considered) in the ensemble. These should be expandable in an eigenstate basis $|u_n\rangle$ (i.e. $A|u_n\rangle = a_n|u_n\rangle$) with $\langle u_n|u_m\rangle = \delta u_{nm}$:

$$\langle \mathbf{A} \rangle = \sum_i p_i \langle \Psi^{(i)}|A|\Psi^{(i)}\rangle = \sum_i \sum_n p_i |\langle u_n|\Psi^{(i)}\rangle|^2 a_n \tag{2.3}$$

with

$$|\Psi^{(i)}\rangle = \sum_n \langle u_n|\Psi^{(i)}\rangle \langle|u_n\rangle = \sum_n c_n^{(i)}|u_n\rangle \tag{2.4}$$

Probabilities appear here twice: once as $|\langle u_n|\Psi^{(i)}\rangle|^2$, which is the probability to find the state $|\Psi^{(i)}\rangle$ in an eigenstate $|u_n\rangle$ of A (with eigenvalue a_n) in the measurement, but also as the probability p_i of finding the ensemble in a quantum mechanical state characterized by $|\Psi^{(i)}\rangle$. By choosing an even more general basis (see e.g. [5]) $|b\rangle$ one can represent the ensemble average more generally as

$$\langle A \rangle = \sum_i p_i \sum_n \sum_m \langle \Psi^{(i)} | b_n \rangle \langle b_n | A | b_m \rangle \langle b_m | \Psi^{(i)} \rangle$$

$$= \sum_n \sum_m \left(\sum_i p_i \langle b_m | \Psi^{(i)} \rangle \langle \Psi^{(i)} | b_n \rangle \right) \langle b_n | \mathbf{A} | b_m \rangle \tag{2.5}$$

The number of terms in the n, m sums is N each while i depends on the composition of the statistical ensemble. In this representation the properties of the ensemble and of the obervable **A** factorize.

References

1. Gerlach, W., Stern, O.: Z. Physik. **9**, 349 (1922)
2. Feynman, R.P., Leighton, R.B., Sands, M.: The Feynman Lectures on Physics, vol. II, chapter 35 and vol. III, chapters 5, 6, 11, 35, Addison-Wesley, Reading (1965)
3. Mitter, H.: Quantentheorie. BI-Hochschultaschenbuch, Mannheim (1976)
4. Mackintosh, A.R.: Eur. J. Phys. **4**, 97 (1983)
5. Sakurai, J.J: Modern Quantum Mechanics. 2nd edn. Addison-Wesley, New York (1994)

Chapter 3
Density Operator, Density Matrix

With the foregoing relations the density operator ρ can be defined:

$$\rho = \sum_i p_i |\Psi^{(i)}\rangle\langle\Psi^{(i)}| \tag{3.1}$$

Its matrix elements are:

$$\langle b_m|\rho|b_n\rangle = \sum_i p_i \langle b_m|\Psi^{(i)}\rangle\langle\Psi^{(i)}|b_n\rangle \tag{3.2}$$

With this definition the ensemble average can be written comfortably:

$$\langle \mathbf{A}\rangle = \sum_n \sum_m \langle b_m|\rho|b_n\rangle\langle b_n|A|b_m\rangle = \mathrm{Tr}(\rho\mathbf{A}) \tag{3.3}$$

Since the trace of a matrix is independent of its different representations $\langle \mathbf{A}\rangle$ can be evaluated in any suitable basis.

3.1 General Properties of ρ

The density matrix (see e.g. [1, 2]) is especially suited for a description of an arbitrary (pure or mixed) polarization state (differently from a wave function). With it averages and expectation values as well as statistical distributions of measurable quantities can be described.

- Every quantized statistical mixture is described exactly and as completely as possible by its density operator.
- Pure and mixed states are being treated in identical ways and operator techniques can be used consistently for the description.
- A wave function $|\Psi\rangle$ can be determined only up to a phase factor $e^{i\phi}$ (which plays no role for the observables). The density matrix ρ, however, is identical for $|\Psi\rangle$ and $|\Psi\rangle e^{i\phi}$.

H. Paetz gen. Schieck, *Nuclear Physics with Polarized Particles*,
Lecture Notes in Physics 842, DOI: 10.1007/978-3-642-24226-7_3,
© Springer-Verlag Berlin Heidelberg 2012

Other properties of ρ:

- The trace of ρ is 1:

$$\text{Tr}(\rho) = \sum_i \sum_n p_i \langle b_n | \Psi^{(i)} \rangle \langle \Psi^{(i)} | b_n \rangle = \sum_i p_i \langle \Psi^{(i)} | \Psi^{(i)} \rangle = 1 \qquad (3.4)$$

(This follows also from $\langle \mathbf{A} \rangle = \text{Tr}(\rho \mathbf{A})$ with $\mathbf{A} = \mathbf{E}$ ($\mathbf{E}$ = unit matrix) and $\sum_k p_k = 1$).

- For $\mathbf{A}$ to have real expectation values ρ must be Hermitean:

$$\rho = \rho^\dagger : \rho_{ik} = \rho_{ki}* \qquad (3.5)$$

- ρ is positively definite (i.e. all diagonal elements are ≥ 0):

$$\langle b_n | \rho | b_n \rangle = \sum_i p_i \langle b_n | \Psi^{(i)} \rangle \langle \Psi^{(i)} | b_n \rangle = \sum_i p_i |\langle b_n | \Psi^{(i)} \rangle|^2 \geq 0 \qquad (3.6)$$

3.2 Distinction Between Pure and Mixed States

3.2.1 Pure State

According to its definition here the density operator is a projection operator:

$$\rho = |\Psi \rangle \langle \Psi| = \sum_n \sum_m |b_n \rangle \langle b_n | \Psi \rangle \langle \Psi | b_m \rangle \langle b_m| \qquad (3.7)$$

in the special basis $|b\rangle$ where $|\Psi\rangle$ is a pure state of the system.

It is *idempotent*, meaning:

$$\rho^2 = \rho \quad \text{or} \quad \rho(\rho - \mathbf{E}) = 0 \qquad (3.8)$$

$$\text{Tr}(\rho^2) = \sum_n \langle u_n | \Psi \rangle \langle \Psi | \Psi \rangle \langle \Psi | u_n \rangle = \sum_n \langle u_n | \Psi \rangle \langle \Psi | u_n \rangle = \text{Tr}(\rho) \qquad (3.9)$$

The eigenvalues of ρ in this case are only 0 or 1. Because of $\text{Tr}(\rho) = 1$: in the diagonal form only a single element of ρ is 1, all others are 0. The structure of ρ therefore is:

$$\rho = \begin{pmatrix} 0 & 0 & \cdot & 0 & 0 \\ 0 & 1 & \cdot & 0 & 0 \\ \cdot & \cdot & \cdot & \cdot & \cdot \\ 0 & 0 & \cdot & 0 & 0 \\ 0 & 0 & \cdot & 0 & 0 \end{pmatrix} \qquad (3.10)$$

As a repetition for this special case: with the eigenfunctions of the system $\langle u_n \rangle$ (often this is shortened to $|n\rangle$, if we deal with a complete ONS of such eigenfunctions; then n only numbers these states and is not a quantum number!) and $\sum_n |u_n\rangle\langle u_n| = \mathbf{E}$ (the completeness relation) the quantum mechanical average (expectation value) of a quantity $\mathbf{A}$ for a pure state is:

$$\langle \mathbf{A} \rangle = \langle \Psi | \mathbf{A} | \Psi \rangle = \sum_n \langle \Psi | n \rangle \langle n | \mathbf{A} | \Psi \rangle = \sum_n |\mathbf{A}| |\langle n | \Psi \rangle|^2 \tag{3.11}$$

$\langle \mathbf{A} \rangle$ is therefore equal to the sum of the values which $\mathbf{A}$ can take, each multiplied with the probability of its appearance. On the other hand,

$$\begin{aligned} \mathrm{Tr}(\rho \mathbf{A}) &= \mathrm{Tr}(|\Psi\rangle\langle\Psi|\mathbf{A}) \\ &= \sum_n \langle n | \Psi \rangle \langle \Psi | \mathbf{A} | n \rangle = \sum_n \langle \Psi | n \rangle \langle n | \mathbf{A} | \Psi \rangle \\ &= \langle \Psi | \mathbf{A} | \Psi \rangle = \langle \mathbf{A} \rangle \end{aligned} \tag{3.12}$$

3.2.2 Mixed State

$\mathrm{Tr}(\rho^2) \leq 1$ is a measure for the degree of mixing of an ensemble (i.e. its deviation from the pure state):

$$\begin{aligned} \mathrm{Tr}(\rho^2) &= \sum_n \sum_i \sum_k p_i p_k \langle n | \Psi^{(i)} \rangle \langle \Psi^{(i)} | \Psi^{(k)} \rangle \langle \Psi^{(k)} | n \rangle \\ &= \sum_i \sum_k p_i p_k |\langle \Psi^{(i)} | \Psi^{(k)} \rangle|^2 \\ &= \sum_k (p_k)^2 \leq \left\{ \left(\sum_k p_k \right)^2 = [\mathrm{Tr}(\rho)]^2 = 1 \right\} \end{aligned} \tag{3.13}$$

where the Schwarz inequality and the following symbolic relations have been used:

$$\mathrm{Tr} \sum = \sum \mathrm{Tr}, \tag{3.14}$$

$$\mathrm{Tr}(ABC) = \mathrm{Tr}(BCA) = \mathrm{Tr}(CAB) \quad \text{and} \tag{3.15}$$

$$\sum |n\rangle\langle n| = 1. \tag{3.16}$$

While the density matrix

$$\rho = \begin{pmatrix} 1 & 0 & 0 \\ 0 & 0 & 0 \\ 0 & 0 & 0 \end{pmatrix} \tag{3.17}$$

represents a pure state (one of complete polarization in the $+x$ direction), the density matrices

$$\rho = 1/2 \begin{pmatrix} 1 & 0 & 0 \\ 0 & 1 & 0 \\ 0 & 0 & 0 \end{pmatrix} \quad \text{or} \quad \rho = 1/3 \begin{pmatrix} 1 & 0 & 0 \\ 0 & 1 & 0 \\ 0 & 0 & 1 \end{pmatrix} \qquad (3.18)$$

are those of mixed (partially polarized) states. The second one—with a uniform occupation of all spin substates—characterizes a completely unpolarized (or maximally mixed) state. All spin substates belonging to spin S have the same weight $p_i = 1/(2S + 1)$ and

$$\rho = \sum_m |m\rangle \frac{1}{2S + 1} \langle m| \quad \text{and} \quad \text{Tr}\left(\rho^2\right) = \frac{1}{2S + 1} < 1 \qquad (3.19)$$

Fano [1] proposed as a general measure for the "degree of polarization" of an ensemble (beam etc.) (historical!):

$$\hat{P} = \left[\text{Tr}(\rho^2) - \frac{1}{2S + 1}\right] \frac{2S + 1}{2S} \qquad (3.20)$$

This quantity is 0 for a completely unpolarized, 1 for a completely polarized ensemble and is valid for arbitrary spin S (as we will see, such a description is generally not sufficient for a partially polarized ensemble!).

3.3 Other General Properties of ρ

• The time development of a physical system with the Hamiltonian H is described by

$$\dot{\rho} = -\frac{i}{\hbar}[H, \rho] \qquad (3.21)$$

• Our "ignorance" about a system can be expressed by defining an "entropy":

$$S = -\sum_k p_k \ln p_k \qquad (3.22)$$

The pure state thus has $S=0$, very "impure" (mixed) states have large positive entropy (another definition is $\langle -k \ln \rho \rangle = -k\text{Tr}(\rho \ln \rho)$).

3.4 Examples for Density Matrices

3.4.1 Spin S = 1/2

As outlined above a pure state is characterized completely by its wave function

$$|\Psi\rangle = a|\chi_{+1/2}\rangle + b|\chi_{-1/2}\rangle \equiv a|\uparrow\rangle + b|\downarrow\rangle = a\begin{pmatrix} 1 \\ 0 \end{pmatrix} + b\begin{pmatrix} 0 \\ 1 \end{pmatrix}. \qquad (3.23)$$

Its density matrix is then

$$\rho = \begin{pmatrix} |a|^2 & ab^* \\ ba^* & |b|^2 \end{pmatrix} \quad \text{with} \quad |a|^2 + |b|^2 = 1 \qquad (3.24)$$

For a single particle with spin 1/2 or, equivalently, for a spin-1/2 beam completely polarized in the direction of the quantization axis (normally the z axis), $a=1$, $b=0$ the situation is depicted in Fig. 3.1 and

$$|\psi\rangle = \begin{pmatrix} 1 \\ 0 \end{pmatrix} \qquad \rho = \begin{pmatrix} 1 & 0 \\ 0 & 0 \end{pmatrix} \qquad (3.25)$$

Similarly in Fig. 3.2 for a particle (beam) completely polarized in the $-z$ direction, $a=0$, $b=1$

$$|\psi\rangle = \begin{pmatrix} 0 \\ 1 \end{pmatrix} \qquad \rho = \begin{pmatrix} 0 & 0 \\ 0 & 1 \end{pmatrix} \qquad (3.26)$$

However, for a single particle (or beam) completely polarized in the general direction β, ϕ with respect to the quantization axis as depicted in Fig. 3.3 has the spinor:

$$\begin{pmatrix} a \\ b \end{pmatrix} = \begin{pmatrix} \cos\frac{\beta}{2} \\ \sin\frac{\beta}{2}e^{i\phi} \end{pmatrix} \qquad (3.27)$$

and consequently the density matrix:

$$\rho = \begin{pmatrix} \cos^2\frac{\beta}{2} & \sin\frac{\beta}{2}\cos\frac{\beta}{2}e^{-i\phi} \\ \sin\frac{\beta}{2}\cos\frac{\beta}{2}e^{i\phi} & \sin^2\frac{\beta}{2} \end{pmatrix}$$
$$= \begin{pmatrix} \rho_{++} & \rho_{+-} \\ \rho_{-+} & \rho_{--} \end{pmatrix} \qquad (3.28)$$

The relation

$$\text{Tr}(\rho) = \text{Tr}(\rho^2) = 1 \qquad (3.29)$$

Fig. 3.1 Spin UP

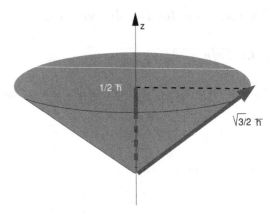

Fig. 3.2 Spin DOWN

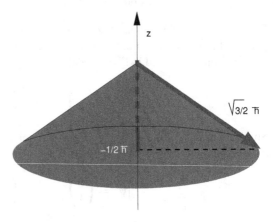

holds. Upon diagonalization (Rotation by $-\beta$; for details of rotations see Chap. 4) this becomes:

$$\rho = \begin{pmatrix} 1 & 0 \\ 0 & 0 \end{pmatrix} \tag{3.30}$$

A completely unpolarized beam with all spin-substates equally occupied has the density matrix:

$$\rho = 1/2 \begin{pmatrix} 1 & 0 \\ 0 & 1 \end{pmatrix} = 1/2 \left[\begin{pmatrix} 1 & 0 \\ 0 & 0 \end{pmatrix} + \begin{pmatrix} 0 & 0 \\ 0 & 1 \end{pmatrix} \right] \tag{3.31}$$

Fig. 3.3 Spin 1/2 in arbitrary direction β relative to z axis, ϕ relative to x axis of a Cartesian coordinate system

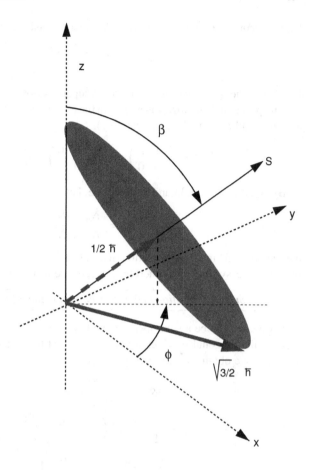

(This corresponds to a superposition of pure spin states with equal weights of 1/2).

Exercise: what sort of state is represented by the density matrix $\rho = 1/2 \begin{pmatrix} 1 & 1 \\ 1 & 1 \end{pmatrix}$?

A general partially polarized beam in an arbitrary direction relative to the z axis is described by a density matrix consisting of two contributions: one contribution p which is completely polarized, and the other $(1-p)$ which is completely unpolarized:

$$\rho = (1 - p)\frac{1}{2} \begin{pmatrix} 1 & 0 \\ 0 & 1 \end{pmatrix} + p \begin{pmatrix} \rho_{++} & \rho_{+-} \\ \rho_{-+} & \rho_{--} \end{pmatrix}$$

$$= \begin{pmatrix} \frac{1-p}{2} + p\cos^2 \frac{\beta}{2} & \frac{p}{2}\sin\beta e^{-i\phi} \\ \frac{p}{2}\sin\beta e^{i\phi} & \frac{1-p}{2} + p\sin^2 \frac{\beta}{2} \end{pmatrix} \tag{3.32}$$

Upon diagonalization (rotation by $-\beta$, unitary transformation) this becomes:

$$\rho = 1/2 \begin{pmatrix} 1+p & 0 \\ 0 & 1-p \end{pmatrix} \tag{3.33}$$

In this case the state can be interpreted as a superposition of the two pure states defined with respect to the quantization axis with the contributions (occupation numbers) $|a|^2$ and $|b|^2$ (and N_+ and N_-, respectively).

$$\rho = N_+ \begin{pmatrix} 1 & 0 \\ 0 & 0 \end{pmatrix} + N_- \begin{pmatrix} 0 & 0 \\ 0 & 1 \end{pmatrix} = \begin{pmatrix} N_+ & 0 \\ 0 & N_- \end{pmatrix}. \tag{3.34}$$

By equating Eqs. 3.33 and 3.34 we see that

$$p = \frac{N_+ - N_-}{N_+ + N_-} \tag{3.35}$$

has the ususal form of a polarization (it is in fact the modulus of the vector polarization of a spin-1/2 system). It is therefore suggestive to introduce the general definition

> "*Spin polarization* is the *expectation value* of a *spin operator*".

If we take $\vec{S}$ to be a spin operator (with vector character such as for the spin-1/2 case) in the diagonal representation with respect to the z direction as quantization axis of ρ one obtains for the z component[1]

$$
\begin{aligned}
p_z = \langle S_z \rangle &= \frac{\text{Tr}(\rho S_z)}{\text{Tr}(\rho)} \\
&= \frac{1}{\text{Tr}(\rho)} \text{Tr}\left[\begin{pmatrix} N_+ & 0 \\ 0 & N_- \end{pmatrix} \begin{pmatrix} 1 & 0 \\ 0 & -1 \end{pmatrix} \right] \\
&= \frac{1}{\text{Tr}(\rho)} \text{Tr}\begin{pmatrix} N_+ & 0 \\ 0 & -N_- \end{pmatrix} \\
&= \frac{N_+ - N_-}{N_+ + N_-}
\end{aligned} \tag{3.36}
$$

which agrees with the usual (naïve) definition of a polarization.

3.4.2 Spin S=1

In the diagonal representation also a naïve definition of the (vector) polarization can be introduced, analogous to the spin-1/2 case:

$$p_z = \frac{N_+ - N_-}{N_+ + N_0 + N_-} \tag{3.37}$$

---

[1]   For the following discussion a description in Cartesian coordinates is assumed.

However, since in this definition no statement about the occupation of the state $|\chi_0\rangle$ has been made, it is evident that at least one additional independent quantity has to be defined in order to be able to describe the spin-1 situation completely. This quantity is called tensor polarization (sometimes also "alignment" for spin-1 particles such as photons as different from vector "polarization") and it is defined as the (normalized) difference between the sum of N_+ and N_- and N_0:

$$P_{zz} = \frac{N_+ + N_- - 2N_0}{N_+ + N_0 + N_-} \tag{3.38}$$

For a pure state, similar to the spin-1/2 case:

$$|\Psi\rangle = a|\chi_1\rangle + b|\chi_0\rangle + c|\chi_{-1}\rangle \equiv a|1\rangle + b|0\rangle + c|-1\rangle = \begin{pmatrix} a \\ b \\ c \end{pmatrix} \tag{3.39}$$

and

$$\rho = \begin{pmatrix} |a|^2 & ab^* & ac^* \\ ba^* & |b|^2 & bc^* \\ ca^* & cb^* & |c|^2 \end{pmatrix} \tag{3.40}$$

For mixed states the ensemble average over the pure states which constitute the ensemble has to be taken.

Examples for spin $S = 1$ The pure states are:

For Spin UP, SIDEWAYS, and DOWN, as shown in Fig. 3.4: the state vectors and density matrices are

$$|\uparrow\rangle = \begin{pmatrix} 1 \\ 0 \\ 0 \end{pmatrix} \qquad |\rightarrow\rangle = \begin{pmatrix} 0 \\ 1 \\ 0 \end{pmatrix} \qquad |\downarrow\rangle = \begin{pmatrix} 0 \\ 0 \\ 1 \end{pmatrix}$$

$$\rho_{+1} = \begin{pmatrix} 1 & 0 & 0 \\ 0 & 0 & 0 \\ 0 & 0 & 0 \end{pmatrix} \quad \rho_0 = \begin{pmatrix} 0 & 0 & 0 \\ 0 & 1 & 0 \\ 0 & 0 & 0 \end{pmatrix} \quad \rho_{-1} = \begin{pmatrix} 0 & 0 & 0 \\ 0 & 0 & 0 \\ 0 & 0 & 1 \end{pmatrix} \tag{3.41}$$

P_z	1	0	-1
P_{zz}	1	-2	1

Mixed states contain the pure states with their respective statistical weights. A completely unpolarized beam is described by

$$\rho = \frac{1}{3} \begin{pmatrix} 1 & 0 & 0 \\ 0 & 1 & 0 \\ 0 & 0 & 1 \end{pmatrix} \tag{3.42}$$

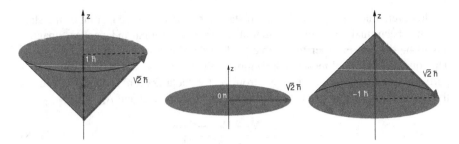

Fig. 3.4 Three possible pure states for spin 1 aligned along the z axis

For a beam with complete positive "alignment" the $+1$ and -1 states are completely occupied and the density matrix is

$$\rho = \frac{1}{2} \begin{pmatrix} 1 & 0 & 0 \\ 0 & 0 & 0 \\ 0 & 0 & 1 \end{pmatrix} \tag{3.43}$$

with $p_z = 0$, $p_{zz} = 1$, $\hat{P} = 1/4$. The situation of maximal negative alignment is obtained when only the 0 state is occupied:

$$\rho = \begin{pmatrix} 0 & 0 & 0 \\ 0 & 1 & 0 \\ 0 & 0 & 0 \end{pmatrix} \tag{3.44}$$

with $p_z = 0$, $p_{zz} = -2$, $\hat{P} = 1$. A case occuring e.g. in the Lambshift polarized-ion source is a mixed state where the $+1$ state has the (relative) occupation number of $2/3$, the 0 state that of $1/3$:

$$\rho = \frac{1}{3} \begin{pmatrix} 2 & \sqrt{2} & 0 \\ \sqrt{2} & 1 & 0 \\ 0 & 0 & 0 \end{pmatrix} \tag{3.45}$$

and $p_z = 2/3$, $p_{zz} = 0$, $\hat{P} = 1/3$. A further case realized there is:

$$\rho = \frac{1}{3} \begin{pmatrix} \frac{1}{2} & 0 & 0 \\ 0 & 2 & 0 \\ 0 & 0 & \frac{1}{2} \end{pmatrix} \tag{3.46}$$

and $p_z = 0$, $p_{zz} = -1$, $\hat{P} = 1/4$.

3.4.3 Rotation of a Pure S=1 State

The rotation of a spin-1/2 state has been indicated above, see Sect. 3.4.1. The general formalism for rotations will be decribed below in Chap. 4. The rotation of ρ by rotation functions is described there in detail.

$$\rho' = \mathbf{D}\rho\mathbf{D}^\dagger \tag{3.47}$$

For $\rho = \begin{pmatrix} 1 & 0 & 0 \\ 0 & 0 & 0 \\ 0 & 0 & 0 \end{pmatrix}$ under a rotation by β, ϕ the rotated density matrix is

$$\rho' = \frac{1}{4} \begin{pmatrix} (1+\cos\beta)^2 & \sqrt{2}(1+\cos\beta)\sin\beta e^{i\phi} & \sin^2\beta e^{2i\phi} \\ \sqrt{2}\sin\beta(1+\cos\beta)e^{-i\phi} & 2\sin^2\beta & \sqrt{2}\sin\beta(1-\cos\beta)e^{i\phi} \\ \sin^2\beta e^{-2i\phi} & \sqrt{2}(1-\cos\beta)\sin\beta e^{-i\phi} & (1-\cos\beta)^2 \end{pmatrix} \tag{3.48}$$

with the rotation matrix (see Chap. 4) for $S=1$:

$$\mathbf{D} = \frac{1}{2} \begin{pmatrix} (1+\cos\beta)e^{i\phi} & -\sqrt{2}\sin\beta e^{i\phi} & (1-\cos\beta)e^{i\phi} \\ \sqrt{2}\sin\beta & 2\cos\beta & -\sqrt{2}\sin\beta \\ (1-\cos\beta)e^{-i\phi} & \sqrt{2}\sin\beta e^{-i\phi} & (1+\cos\beta)e^{-i\phi} \end{pmatrix} \tag{3.49}$$

ρ' has the trace of 1 and represents—which is not easily recognizable from its external form—also a pure state.

3.5 Complete Description of Spin Systems

The number of parameters for a complete description of a system with spin depends on the value of S. For example for $S=1$: besides the intensity (or number of particles) being a scalar quantity or a tensor of rank 0, the vector polarization (rank 1) needs three, the tensor polarization (a tensor of rank 2) needs nine components minus one for a normalization condition (making it traceless). Altogether these are 11 parameters (12 with the intensity). The polarized beam e.g. from an ion source has rotational symmetry around the z axis, the direction of the beam. Therefore the tensor will be symmetric, i.e. $p_{xx} = p_{yy}$, $p_{xy} = p_{yx}$ etc., which reduces the number of parameters to 8 (9). For the polarization of particles produced in a nuclear reaction this reduction does not hold. But symmetries like parity conservation help reduce the number of observables . A special case are again pure states. Such a state is described by $4S = 2(2S+1) - 2$ real parameters ($2S+1$ complex numbers minus one normalization condition minus one common phase). Normally for mixed states the number of parameters is $4S(S+1)$ $(2(2S+1)^2$ real numbers minus $(2S+1)^2$ hermitecity conditions minus one normalization: $2(2S+1)^2 - (2S+1)^2 - 1 = 4S(S+1)$. In numbers:

	Pure state in z direction	Pure state in arbitrary direction β, ϕ relative to x,y,z	Mixed state general
$S = 1/2$	1 p_z	2 e.g. β, ϕ	3 e.g. p_x, p_y, p_z
$S = 1$	2 e.g. p_z, p_{zz}	4 e.g. p_z, p_{zz} β, ϕ	8 e.g. p_x, p_y, p_z $p_{yy}, p_{zz}, p_{xz}, p_{xy}, p_{yz}$
$S = 3/2$	3	6	15

$$(3.50)$$

3.6 Expansions of the Density Matrix, Spin Tensor Moments

The density matrix directly is not well suited to describe observables . It is used to calculate expectation values of operators. The intensity of the incoming or outgoing particles in a nuclear reaction, and also the number of target particles are proportional to the trace of the relevant density matrix ρ. Since these quantities transform as scalars they are, up to a normalization, equal to $1 = \text{Tr}(\rho \mathbf{E})$. It is customary to normalize the incident density matrix exactly to one $(\text{Tr}(\rho)_{in} = 1)$ which automatically leads to $\text{Tr}(\rho_{fin}) \neq 1$. The expansion of the density matrix into basis systems with well-defined properties (e.g. under transformations like rotations) provides a description of observables with corresponding behavior. Another requirement of the definition is certainly the correspondence of the so-defined quantities with older naïvely defined quantities. Rank-1 polarization must behave like a vector with a maximum value of 1, rank-2 polarization like a rank-2 tensor etc.

Therefore the prescription is: expansion of the density matrix into a complete set of *orthogonal* basis matrices with the desired properties and with the expansion coefficients being the new parameter set:

$$\rho = \sum_j \lambda_j U_j^\dagger \tag{3.51}$$

For the U only orthogonality is required: $\text{Tr}(U_i U_k^\dagger) = \delta_{ik}(2S + 1)$, but not necessarily hermitecity since in general the λ_j are complex. Because ρ has $(2S + 1)^2$ complex elements $(2S + 1)^2$ basis matrices are needed (e.g. four for $S=1/2$). The expansion runs from $j=1$ to $(2S+1)^2$. Every tensor of rank k has $2k+1$ components, therefore:

$$(2S + 1)^2 = \sum_{k=0}^{k_{max}} (2k + 1) = \frac{1}{2}(k_{max} + 1)(2k_{max} + 2) = (k_{max} + 1)^2 \quad (3.52)$$

Thus, the maximum rank of spin tensors necessary for a complete description of a spin system is $k_{max} = 2S$.

For the interpretation of the expansion coefficients λ_j: multiply both sides with U_i, form the trace, use orthogonality and the definition of the expectation value of an operator:

$$\mathrm{Tr}(\rho U_i) = \mathrm{Tr}(U_i \rho) = \mathrm{Tr}\left(\sum_{j=1}^{(2S+1)^2} \lambda_j U_i U_j^\dagger\right) = \sum_j \lambda_j \mathrm{Tr}\left(U_i U_j^\dagger\right)$$

$$= \sum_j \lambda_j (2S + 1)\delta_{ij} = (2S + 1)\lambda_i \equiv \langle U_i \rangle \quad (3.53)$$

where the average is to be taken over the ensemble (e.g. the beam). By comparison one obtains:

$$\lambda_i = \frac{1}{2S + 1} \langle U_i \rangle_{beam} \quad (3.54)$$

and

$$\rho = \frac{1}{2S + 1} \sum_{j=1}^{(2S+1)^2} \langle U_j \rangle_{beam} U_j^\dagger \quad (3.55)$$

Like in other expansions (electromagnetic, mass moments...) the coefficients $\langle U_j \rangle$ are called *moments* of the expansion, here: *(spin) tensor moments*. This concept was introduced by Fano [3].

The most important of such expansions are those into Cartesian and into spherical tensors. The latter ones behave under rotations like spherical harmonics $Y_\ell^m(\Theta, \Phi)$. It is useful to choose irreducible representations for the U_j. All transformations, e.g. rotations are then linear and different ranks of submatrices will not be mixed by the transformation, but only transform within their rank (components of the tensor polarization do not produce a vector polarization).

3.6.1 Expansion of ρ in a Cartesian Basis for Spin $S = 1/2$

$$|\chi\rangle = \begin{pmatrix} a \\ b \end{pmatrix}$$

$$\rho = \begin{pmatrix} \overline{|a|^2} & \overline{ab^*} \\ \overline{ba^*} & \overline{|b|^2} \end{pmatrix} \quad (3.56)$$

As basis matrices U_i the unit matrix and the three Pauli matrices are chosen: $\vec{\sigma}$:

$$U_1 = \mathbf{E} = \begin{pmatrix} 1 & 0 \\ 0 & 1 \end{pmatrix}$$

$$U_2 = \sigma_x = \begin{pmatrix} 0 & 1 \\ 1 & 0 \end{pmatrix}$$

$$U_3 = \sigma_y = \begin{pmatrix} 0 & -i \\ i & 0 \end{pmatrix}$$

$$U_4 = \sigma_z = \begin{pmatrix} 1 & 0 \\ 0 & -1 \end{pmatrix} \tag{3.57}$$

This leads to:

$$\rho = \frac{1}{2}(1 + \langle \sigma \rangle \sigma) = \frac{1}{2}(1 + \vec{p}\vec{\sigma})$$

$$= 1/2 \begin{pmatrix} 1 + p_z & p_x - ip_y \\ p_x + ip_y & 1 - p_z \end{pmatrix} \tag{3.58}$$

By comparing coefficients:

$$p_x = 2\mathrm{Re}(\overline{ba^*}); \quad p_y = 2\mathrm{Im}(\overline{ba^*}); \quad p_z = \overline{|a|^2} - \overline{|b|^2} \tag{3.59}$$

The result for p_z corresponds to the naïve definition of a polarization!

3.6.2 Spin S=1

Here one needs $(2S + 1)^2 = 9$ basis matrices of rank 2. They are obtained applying the direct product of tensors with rank 1 (vectors). Because of their rotation properties the three Cartesian Pauli matrices for $S=1$ and the unit matrix are chosen from which one can form 13 3×3 matrices of rank 2. Beginning with

$$\mathbf{E} = \begin{pmatrix} 1 & 0 & 0 \\ 0 & 1 & 0 \\ 0 & 0 & 1 \end{pmatrix} \quad S_x = \frac{1}{\sqrt{2}} \begin{pmatrix} 0 & 1 & 0 \\ 1 & 0 & 1 \\ 0 & 1 & 0 \end{pmatrix}$$

$$S_y = \frac{1}{\sqrt{2}} \begin{pmatrix} 0 & -i & 0 \\ i & 0 & -i \\ 0 & i & 0 \end{pmatrix} \quad S_z = \begin{pmatrix} 1 & 0 & 0 \\ 0 & 0 & 0 \\ 0 & 0 & -1 \end{pmatrix} \tag{3.60}$$

one forms:

$$S_x^2 = \frac{1}{2}\begin{pmatrix} 1 & 0 & 1 \\ 0 & 2 & 0 \\ 1 & 0 & 1 \end{pmatrix} \qquad S_y^2 = \frac{1}{2}\begin{pmatrix} 1 & 0 & -1 \\ 0 & 2 & 0 \\ -1 & 0 & 1 \end{pmatrix}$$

$$S_z^2 = \begin{pmatrix} 1 & 0 & 0 \\ 0 & 0 & 0 \\ 0 & 0 & 1 \end{pmatrix} \qquad S_x S_y = \frac{1}{2}\begin{pmatrix} i & 0 & -i \\ 0 & 0 & 0 \\ i & 0 & -i \end{pmatrix}$$

$$S_x S_z = \frac{1}{\sqrt{2}}\begin{pmatrix} 0 & 0 & 0 \\ 1 & 0 & -1 \\ 0 & 0 & 0 \end{pmatrix} \qquad S_y S_z = \frac{1}{\sqrt{2}}\begin{pmatrix} 0 & 0 & 0 \\ i & 0 & i \\ 0 & 0 & 0 \end{pmatrix} \qquad (3.61)$$

and similarly for $S_y S_x$, $S_z S_x$ and $S_z S_y$. The S_i are hermitean: $S_i S_j = (S_j S_i)^\dagger$, i.e. the $S_i S_j$ are connected with the $S_j S_i$ via commutation relations. The antisymmetric combinations

$$S_i S_j - S_j S_i = i S_k \qquad (3.62)$$

and the symmetric combinations

$$S_i S_j + S_j S_i = (S_i S_j)^\dagger + (S_j S_i)^\dagger = (S_i S_j + S_j S_i)^\dagger \qquad (3.63)$$

are automatically hermitean. In a decomposition

$$S_i S_j = \frac{1}{2}(S_i S_j + S_j S_i) + \frac{1}{2}(S_i S_j - S_j S_i) \qquad (3.64)$$

into a symmetric and an antisymmetric part the latter provides nothing new because of the commutation relations. Therefore only the symmetric part is kept to form the new combinations:

$$S_{ij}' = \frac{1}{2}(S_i S_j + S_j S_i) = 1/2(S_i S_j + (S_i S_j)^\dagger) \qquad (3.65)$$

Of these there are exactly six. Five are needed, which allows to introduce a physical condition into the final definition of the S_{ij}: The polarization of the unpolarized ensemble ought to be zero: $(\mathrm{Tr}(S_j) = \mathrm{Tr}(S_{ij}') = 0)$. Since this is not fulfilled for the S_{ij}' with $i=j$, a new definition is used

$$S_{ij} =: 3S_{ij}' - 2\delta_{ij}\mathbf{E} = 3/2\,(S_i S_j + S_j S_i) - 2\delta_{ij}\mathbf{E}, \qquad (3.66)$$

leading to: $S_{xx} + S_{yy} + S_{zz} = 0$, $\mathrm{Tr}\,(S_{ij}) = \mathrm{Tr}\,(S_i) = 0$. All "polarizations" of the unpolarized ensemble are then zero. Of the three diagonal elements S_{ii} only two are independent. Also, since no two S_{ii} are orthogonal, one uses two orthogonal linear combinations of the S_{ii} instead. Several combinations are possible, e.g. $(S_{xx} - S_{yy}; S_{zz})$ or $(S_{zz} - S_{xx}; S_{yy})$. Here $S_{xx} + S_{yy}$ and $S_{xx} - S_{yy}$ will be chosen. With $\mathrm{Tr}(S_i^2) = 2$, $\mathrm{Tr}(S_{ij}^2) = 9/2$ and $\mathrm{Tr}(S_{xx} + S_{yy})^2 = 6$, $\mathrm{Tr}(S_{xx} - S_{yy})^2 = 18$ the Cartesian expansion basis is:

$$U_1 = \mathbf{E};$$

$$U_2 = \sqrt{3/2}S_x; \quad U_3 - \sqrt{3/2}S_y; \quad U_4 = \sqrt{3/2}S_z;$$

$$U_5 = \sqrt{2/3}S_{xy}; \quad U_6 = \sqrt{2/3}S_{xz}; \quad U_7 = \sqrt{2/3}S_{yz};$$

$$U_8 = \sqrt{1/2}(S_{xx} + S_{yy});$$

$$U_9 = \sqrt{1/6}(S_{xx} - S_{yy}). \tag{3.67}$$

With $\langle S_{xx} \rangle + \langle S_{yy} \rangle + \langle S_{zz} \rangle = 0$ and $S_{xx} + S_{yy} + S_{zz} = 0$ one obtains

$$\rho = \frac{1}{3} \sum_{k=1}^{(2S+1)^2} \langle U_k \rangle U_k^\dagger$$

$$= \frac{1}{3} \left(\mathbf{E} + \frac{3}{2} \sum_i \langle S_i \rangle S_i + \frac{1}{3} \sum_{ij} \langle S_{ij} \rangle S_{ij} \right) \tag{3.68}$$

(for a different derivation see [4]). The sum over ij is meant such that for $i=j$ one, for $i \neq j$ two terms appear). Written out:

$$\rho =$$

$$\frac{1}{3} \begin{pmatrix} 1 + \frac{3}{2}p_z + \frac{1}{2}p_{zz} & \frac{1}{\sqrt{2}}\left[\frac{3}{2}(p_x - ip_y) + (p_{xz} - ip_{yz})\right] & \frac{1}{2}(p_{xx} - p_{yy}) - ip_{xy} \\ \frac{1}{\sqrt{2}}\left[\frac{3}{2}(p_x + ip_y) + (p_{xz} + ip_{yz})\right] & 1 - p_{zz} & \frac{1}{\sqrt{2}}\left[\frac{3}{2}(p_x - ip_y) - (p_{xz} - ip_{yz})\right] \\ \frac{1}{2}(p_{xx} - p_{yy}) + ip_{xy} & \frac{1}{\sqrt{2}}\left[\frac{3}{2}(p_x + ip_y) - (p_{xz} + ip_{yz})\right] & 1 - \frac{3}{2}p_z + \frac{1}{2}p_{zz} \end{pmatrix}$$

$$\tag{3.69}$$

The polarization parameters of rank k ($k=0, 1, 2$) form tensors of rank k. For $k=0$ it is a scalar proportional to some intensity, for $k=1$ it is a vector with three components (the vector polarization) and for $k=2$ it is a tensor of rank 2, represented by a symmetric traceless $(k+1) \times (k+1)$ (thus 3×3) matrix of the form

$$\begin{pmatrix} p_{xx} & p_{xy} & p_{xz} \\ p_{xy} & p_{yy} & p_{zy} \\ p_{xz} & p_{zy} & p_{zz} \end{pmatrix} \tag{3.70}$$

with five independent parameters. Taken together the spin-1 system has nine parameters.

The relation to the components a, b, c of an ($S=1$) spinor is:

$$p_x = \sqrt{2}\mathrm{Re}\,(\overline{ba^*} + \overline{cb^*}); \quad p_y = \sqrt{2}\mathrm{Im}\,(\overline{ba^*} + \overline{cb^*}); \quad p_z = \overline{|a|^2} - \overline{|c|^2}$$

$$p_{xz} = \frac{1}{\sqrt{2}}\mathrm{Re}\,(\overline{ba^*} - \overline{cb^*}); \quad p_{yz} = \frac{1}{\sqrt{2}}\mathrm{Im}\,(\overline{ba^*} - \overline{cb^*}); \quad p_{xy} = \mathrm{Im}\,(\overline{ca^*})$$

$$p_{zz} = -(p_{xx} + p_{yy}) = 1 - 3\overline{|b|^2} = \overline{|a|^2} + \overline{|c|^2} - 2\overline{|b|^2}$$

$$\tag{3.71}$$

in agreement with the naïve definitions of p_z and p_{zz}.

3.6.3 Limiting Values of the Polarization Components

Because of $\text{Tr}(\rho) = \overline{|a|^2} + \overline{|b|^2} + \overline{|c|^2} = 1$

$$\overline{|a|^2} + \overline{|c|^2} \leq 1 \tag{3.72}$$

holds and thus:

$$\left| \overline{|a|^2} - \overline{|c|^2} \right| \leq \overline{|a|^2} + \overline{|c|^2} \leq 1 \tag{3.73}$$

From this follows:

$$|p_z| \leq 1 \quad \text{or} \quad -1 \leq p_z \leq +1 \tag{3.74}$$

and, because the choice of axes is arbitrary:

$$|p_x| \leq 1 \quad \text{and} \quad |p_y| \leq 1. \tag{3.75}$$

In the same way it can be shown that

$$|p_{xz}| \leq 3/2; \quad |p_{yz}| \leq 3/2; \quad |p_{xy}| \leq 3/2; \quad |p_{xx} - p_{yy}| \leq 3. \tag{3.76}$$

and therefore:

$$-2 \leq p_{zz} \leq +1 \tag{3.77}$$

The components of the vector and tensor polarization are not independent of each other but are related via the occupation numbers (or the occupation probabilities) N_i of the three substates of the spin-1 system. In the diagonal representation ($\overline{|a|^2} = N_+$; $\overline{|b|^2} = N_0$; $\overline{|c|^2} = N_-$) the normalization with $N_+ + N_0 + N_- = 1$ holds together with the "positivity condition" $N_i \geq 0$. This is sufficient to determine the values of the vector and tensor polarization as functions of the occupation numbers (Fig. 3.5).

3.6.4 Expansion Into Spherical Tensors

The representation of every vector can be Cartesian or spherical. We begin with the position vector as an example: the spherical components of a position vector

$$\mathbf{r} = \begin{pmatrix} x \\ y \\ z \end{pmatrix} \text{ are:}$$

$$r_{\pm 1} = \mp \frac{1}{\sqrt{2}} (x \pm iy) \quad \text{and} \quad r_0 = z \tag{3.78}$$

Fig. 3.5 Range of values of the polarization components of a spin-1 system as functions of the (relative) occupation numbers of a spin-1 system with additional conditions $N_+ + N_0 + N_- = 1$ and $N_i \geq 0$

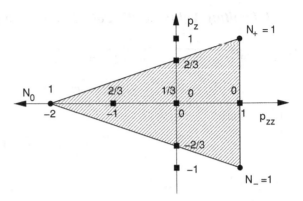

The corresponding so-called "spatial harmonic" is defined as:

$$\mathcal{Y}_{1m}(\mathbf{r}) = \sqrt{3/(4\pi)}\, r_m = r^1 Y_1^m(\Theta, \Phi), \tag{3.79}$$

thus

$$r Y_1^{\pm 1}(\Theta, \Phi) = \sqrt{3/(4\pi)} \left(\mp \frac{1}{\sqrt{2}}(x \pm iy) \right) \tag{3.80}$$

and

$$r Y_1^0(\Theta, \Phi) = \sqrt{3/(4\pi)}\, z \tag{3.81}$$

It describes the position (x, y, z) in spherical polar coordinates (r, Θ, Φ). Analogously every quantity transforming as a vector can be represented spherically.

Since tensors (tensor operators) of arbitrary rank can be generated by the combination (the direct or external product) from the components of a given set of vectors (vector operators) the spherical tensors τ_{kq} (e.g. of rank $k=2$ or higher) can be contructed from spin operators $\vec{S}$ in their spherical representation τ_{kq} (for $k=1$). The resulting product operators in general are *reducible*, i.e. their transformation properties e.g. under rotations will not be simple. By forming the direct product of two vectors in three dimensions, a (Cartesian) tensor with nine elements will result which behave differently under rotations, namely like a scalar (the trace of this tensor), like a vector (whose components are the three cross products of the original vector components), and like a tensor of rank 2.

Reducible tensors can be decomposed into objects (tensors) which still behave differently but *independently* e.g. under rotations. More formally: a reducible Cartesian tensor T_{ij} of rank 2 can be generated from two vectors $\mathbf{U}$ and $\mathbf{V}$ using the prescription $\mathbf{T} = \mathbf{U} \otimes \mathbf{V}$, i.e. $T_{ij} \equiv U_i V_j$. However, it can be decomposed (reduced) in the following way:

$$U_i V_j = 1/3 \mathbf{UV} \delta_{ij} + 1/2(U_i V_j - U_j V_i) + [1/2(U_i V_j + U_j V_i) - 1/3 \mathbf{UV} \delta_{ij}] \tag{3.82}$$

Thus the tensor is decomposed into a scalar (tensor of rank 0), an antisymmetric tensor (vector or cross product), which transforms like a vector (tensor of rank 1), and a 3×3 traceless, symmetric tensor of rank 2. Symbolically:

$$3 \otimes 3 = 1 \oplus 3 \oplus 5 \qquad (3.83)$$

This, however, is just a decomposition of the reducible Cartesian tensor into spherical components (with rotation properties like the spherical harmonics of ranks 0, 1 and 2). The matrices describing the tensor can then be decomposed into submatrices along the main diagonal which transform only linearly according to their rank and without influencing the other submatrices. This "pure" behavior under rotations is displayed only by the special linear combinations of the tensor components defined above, but not by these alone. Under rotations of systems represented by reducible tensors all components would have to be transformed in common and according to the usual transformation rules for tensors—namely non-linearly in the rotation functions.

Especially for higher spins and generally because of these transformation properties the use of irreducible *spherical tensors* is preferred for the description of polarization. Generalizing this to arbitrary spin systems: They transform under rotations *linearly in the rotation functions* like the angular momentum eigenfunctions, the spherical harmonics $Y_\ell^m(\Theta, \Phi) = P_\ell^m(\Theta)e^{im\Phi}$.

These tensors are generated by a special combination of spin vector operators (which are irreducible per se!), by applying the principles of vector coupling of angular momentum operators using Clebsch–Gordan coefficients where operators of higher rank with identical transformation properties as those of the constituents are generated:

$$\tau_{KQ} = \sum_{q,q'} \left(kk'qq' | K Q\right) \tau_{kq}\tau_{k'q'} \qquad (3.84)$$

The density matrices of higher-spin systems can be expanded into a complete set of such basis matrices—just like for spin-1/2 systems. We define as a "spherical basis" the irreducible tensors of rank K in a spherical coordinate representation (short: "spherical tensors"). They have to fulfill at least one condition: τ_{KQ} has to be transformed like the spherical harmonic of rank K and component Q Y_K^Q under spatial rotations. The spherical tensors, like the spherical harmonics, are irreducible, i.e. under transformations (rotations) they will always be transformed into tensors of equal rank (about rotations see Chap. 4) with:

$$\tau_{KQ'} = \sum_Q D_{Q'Q}^K(\alpha, \beta, \gamma)\tau_{KQ} \qquad (3.85)$$

3.6.5 Example for the Construction of a Set of Spherical Tensors for S=1

We choose as basis:

$$\mathbf{E} = \begin{pmatrix} 1 & 0 & 0 \\ 0 & 1 & 0 \\ 0 & 0 & 1 \end{pmatrix}$$

$$S_1 = -\frac{1}{\sqrt{2}}(S_x + iS_y) = \begin{pmatrix} 0 & -1 & 0 \\ 0 & 0 & -1 \\ 0 & 0 & 0 \end{pmatrix}$$

$$S_{-1} = \frac{1}{\sqrt{2}}(S_x - iS_y) = \begin{pmatrix} 0 & 0 & 0 \\ 1 & 0 & 0 \\ 0 & 1 & 0 \end{pmatrix}$$

$$S_0 \equiv S_z = \begin{pmatrix} 1 & 0 & 0 \\ 0 & 0 & 0 \\ 0 & 0 & -1 \end{pmatrix} \tag{3.86}$$

These basis matrices are non-hermitean: $S_{\pm 1}^\dagger = -S_{\mp 1}$ and:

$$\text{Tr}\left(S_1 S_1^\dagger\right) = \text{Tr}\left(S_{-1} S_{-1}^\dagger\right) = \text{Tr}\left(S_0 S_0^\dagger\right) = 2 \tag{3.87}$$

From these one basis tensor of rank 0, three of rank 1, and five of rank 2 are constructed in analogy to the procedure with spherical harmonics (which are by definition spherical tensors) for which holds:

$$Y_K^Q = \sum_{q_1 q_2}(k_1 k_2 q_1 q_2 | K Q) Y_{k_1}^{q_1} Y_{k_2}^{q_2} \tag{3.88}$$

e.g. for K=2:

$$Y_2^Q = \sum_{q=-1}^{+1} (11qQ - q|2Q) Y_1^q Y_1^{Q-q}. \tag{3.89}$$

Similarly tensor operators of still higher rank can be constructed by coupling of tensor operators of lower rank ("contraction"). As spin operators for $S=1$ the four operators defined above $\mathbf{E}$, S_0 and $S_{\pm 1}$ can be used to construct the missing operators of rank 2 S_{2Q}

$$S_{2Q} = \sum_{m=-1}^{1} (11mQ - m|2Q) S_m S_{Q-m}. \tag{3.90}$$

Thus, e.g. (with $Q=0$):

$$S_{20} = \sum_m (11mQ - m|20) S_m S_{Q-m} = 1/\sqrt{6}\,(S_1 S_{-1} + S_{-1} S_1) + \sqrt{2/3}\,S_0^2$$

$$= 1/\sqrt{6}(S_1 S_1^\dagger + S_{-1} S_{-1}^\dagger) + \sqrt{2/3}\,S_0^2$$

$$= -1/\sqrt{6}\left(S^2 - S_0^2\right) + \sqrt{2/3}\,S_0^2 = 1/\sqrt{6}\left(-2 + 3S_0^2\right)$$

$$(3.91)$$

(since S^2 is diagonal with $S(S+1) = 2$). This procedure can be continued. All one has to know are the Clebsch–Gordan vector coupling coefficients. In this way for $S = 1$ the following irreducible spherical tensors (normalized according to Lakin [5], i.e. following the Madison convention) are obtained:

$$\tau_{00} = \mathbf{E} = \begin{pmatrix} 1 & 0 & 0 \\ 0 & 1 & 0 \\ 0 & 0 & 1 \end{pmatrix}$$

$$\tau_{11} = \sqrt{3/2}\,S_{+1} = \sqrt{3/2}\begin{pmatrix} 0 & -1 & 0 \\ 0 & 0 & -1 \\ 0 & 0 & 0 \end{pmatrix}$$

$$\tau_{1-1} = \sqrt{3/2}\,S_{-1} = \sqrt{3/2}\begin{pmatrix} 0 & 0 & 0 \\ 1 & 0 & 0 \\ 0 & 1 & 0 \end{pmatrix}$$

$$\tau_{10} = \sqrt{3/2}\,S_0 = \sqrt{3/2}\begin{pmatrix} 1 & 0 & 0 \\ 0 & 0 & 0 \\ 0 & 0 & -1 \end{pmatrix}$$

$$\tau_{22} = \sqrt{3}\,S_{22} = \sqrt{3}\,S_{+1}^2 = \sqrt{3}\begin{pmatrix} 0 & 0 & 1 \\ 0 & 0 & 0 \\ 0 & 0 & 0 \end{pmatrix}$$

$$\tau_{21} = \sqrt{3}\,S_{21} = \sqrt{3/2}(S_0 S_1 + S_1 S_0) = \sqrt{3/2}\begin{pmatrix} 0 & -1 & 0 \\ 0 & 0 & 1 \\ 0 & 0 & 0 \end{pmatrix}$$

$$\tau_{20} = \sqrt{3}\,S_{20} = 1/\sqrt{2}(3S_0^2 - 2) = 1/\sqrt{2}\begin{pmatrix} 1 & 0 & 0 \\ 0 & -2 & 0 \\ 0 & 0 & 1 \end{pmatrix} \qquad (3.92)$$

The missing components are calculated with

$$\tau_{K-Q} = (-)^Q \tau_{KQ}^\dagger \qquad (3.93)$$

For $S = \frac{3}{2}$ spherical tensors are constructed similarly:

$$S_{3Q} = \sum_{m=-1}^{1} (21mQ - m|3Q) S_{2m} S_{1,Q-m}. \qquad (3.94)$$

3.6.6 Spin Tensor Moments

The quantities which specify the polarization in the spherical representation are the expectation values of these tensor operators, the so-called *spin tensor moments* $t_{KQ} = \langle \tau_{KQ} \rangle$.

The hermitecity of ρ entails the hermitecity of the t_{KQ}: $t_{K-Q} = (-)^{Q} t_{KQ}^{*}$. Therefore the density matrix can be expressed in terms of tensor moments.

3.6.7 Spherical Tensors , Density Matrix, and Tensor Moments for Spin S = 1/2

The irreducible basis (as linear combinations of Pauli operators the spin tensors are in this case automatically *irreducible*)

$$\tau_{00} = \begin{pmatrix} 1 & 0 \\ 0 & 1 \end{pmatrix}$$

$$\tau_{10} = \begin{pmatrix} 1 & 0 \\ 0 & -1 \end{pmatrix} = \sigma_z$$

$$\tau_{11} = -\sqrt{2} \begin{pmatrix} 0 & 1 \\ 0 & 0 \end{pmatrix} = -1/\sqrt{2}(\sigma_x + i\sigma_y)$$

$$\tau_{1-1} = \sqrt{2} \begin{pmatrix} 0 & 0 \\ 1 & 0 \end{pmatrix} = 1/\sqrt{2}(\sigma_x - i\sigma_y) \qquad (3.95)$$

The expansion of the density matrix is

$$\rho = 1/2 \sum_{KQ} \langle \tau_{KQ} \rangle \tau_{KQ}^{\dagger} \qquad (3.96)$$

and

$$\mathrm{Tr}(\tau_{KQ} \tau_{K'Q'}^{\dagger}) = 2\delta_{KQ,K'Q'} \qquad (3.97)$$

and therefore

$$\rho = \frac{1}{2}\begin{pmatrix} 1+t_{10} & \sqrt{2}t_{1-1} \\ -\sqrt{2}t_{11} & 1-t_{10} \end{pmatrix} = \frac{1}{2}\begin{pmatrix} 1+p_z & p_x - ip_y \\ p_x + ip_y & 1-p_z \end{pmatrix} \tag{3.98}$$

By comparison:

$$t_{00} \quad \text{is proportional to an intensity}$$
$$t_{10} = p_z \tag{3.99}$$
$$t_{1\pm1} = \mp 1/\sqrt{2}(p_x \pm ip_y).$$

3.6.8 Density Matrix and Tensor Moments for Spin S = 1

$$\rho = \frac{1}{3}\begin{pmatrix} 1+\sqrt{\frac{3}{2}}t_{10} + \frac{1}{\sqrt{2}}t_{20} & \sqrt{\frac{3}{2}}(t_{1-1}+t_{2-1}) & \sqrt{3}t_{2-2} \\ -\sqrt{\frac{3}{2}}(t_{11}+t_{21}) & 1-\sqrt{2}t_{20} & \sqrt{\frac{3}{2}}(t_{1-1}-t_{2-1}) \\ \sqrt{3}t_{22} & -\sqrt{\frac{3}{2}}(t_{11}-t_{21}) & 1-\sqrt{\frac{3}{2}}t_{10} + \frac{1}{\sqrt{2}}t_{20} \end{pmatrix} \tag{3.100}$$

By comparison with Eqs. 3.40 and 3.69 the connection between the t_{kq}, the wavefunction amplitudes, and the Cartesion tensor components is obtained: tensor moments of *rank k = 0* are proportional to an intensity (one scalar, invariant under rotations):

$$t_{00} = 1 = \overline{|a|^2} + \overline{|b|^2} + \overline{|c|^2}. \tag{3.101}$$

Tensor moments of *rank k = 1* describe the vector polarization (three components, transformation properties of a vector):

$$t_{1\pm1} = -\sqrt{3/2}\ (\overline{ba^*} + \overline{cb^*}) = \mp\sqrt{3}/2\ (p_x \pm ip_y)$$
$$t_{10} = \sqrt{3/2}\ (\overline{|a|^2} - \overline{|c|^2}) = \sqrt{3/2}\ p_z \tag{3.102}$$

Tensor moments of *rank k = 2* describe the tensor polarization (eight independent components, transformation properties of a second-rank tensor):

$$t_{20} = 1/\sqrt{2}\ (\overline{|a|^2} + \overline{|c|^2} - 2\overline{|b|^2}) = 1/\sqrt{2}\ (1 - 3\overline{|b|^2}) = 1/\sqrt{2}\ p_{zz}$$
$$t_{2\pm1} = \sqrt{3/2}\ (\overline{cb^*} - \overline{ba^*}) = \mp 1/\sqrt{3}\ (p_{xz} \pm ip_{yz})$$
$$t_{2\pm2} = \sqrt{3}\ \overline{ca^*} = 1/\left(2\sqrt{3}\right)\ (p_{xx} - p_{yy} \pm 2ip_{xy}) \tag{3.103}$$

Inversely:

$$p_x = -1/\sqrt{3} \, (t_{11} - t_{1-1})$$

$$p_y = i/\sqrt{3} \, (t_{11} + t_{1-1})$$

$$p_z = \sqrt{2/3} \, t_{10}$$

$$p_{xx} = \sqrt{3}/2 \, (t_{22} + t_{2-2}) - 1/\sqrt{2} \, t_{20}$$

$$p_{yy} = -\sqrt{3}/2 \, (t_{22} + t_{2-2}) - 1/\sqrt{2} \, t_{20}$$

$$p_{zz} = \sqrt{2} \, t_{20}$$

$$p_{xy} = p_{yx} = -i\sqrt{3}/2 \, (t_{22} - t_{2-2})$$

$$p_{xz} = p_{zx} = -\sqrt{3}/2 \, (t_{21} - t_{2-1})$$

$$p_{yz} = p_{zy} = i\sqrt{3}/2 \, (t_{21} + t_{2-1}) \tag{3.104}$$

For analyzing powers of nuclear reactions the same relations apply (with upper-case variables T_{kq} and A_i or A_{ik}).

In analogy to other expansions into momenta in physics (e.g. mass distributions ($\rightarrow$ moment of inertia etc.), multipole expansions of electric charge or current distributions etc.) the coefficients of expansions into tensor moments have a geometrical interpretation. In the spherical representation, where the angular dependences are those of the spherical harmonics $Y_\ell^m(\Theta, \Phi)$ this can be visualized, see [6], for tensor moments with ranks ≥ 2. The anisotropy, produced by the polarization of an ensemble (an unpolarized ensemble will not have a preferred direction and is represented by a sphere) is symbolized by hypersurfaces described by a radius vector of length

$$R^{(k)}(\Theta, \Phi) = R_0 \left[1 + \sum_q t_{kq} Y_k^{q*}(\Theta, \Phi) \right]. \tag{3.105}$$

For tensor-polarized spin-1 particles (rank $k=2$) and e.g. one component $q=0$ this is reduced to

$$R^{(2)}(\Theta, \Phi = 0) = R_0 \left[1 + t_{20} Y_2^q(\Theta, \Phi = 0) \right]$$

$$= R_0 \left[1 + t_{20} \sqrt{\frac{5}{4\pi}} P_2(\cos \Theta) \right]. \tag{3.106}$$

Figure 3.6 shows the spatial anisotropy produced by the pure tensor polarization component t_{20} of an ensemble completely polarized in the z direction in the two limiting cases $t_{20} = +1/\sqrt{2}$ and $-\sqrt{2}$. The length of the radius vector R is a measure of the probability of measuring this tensor-polarization component in the corresponding direction, or expressed differently: under a rotation of this polarization state a tensor-polarization analyzer would measure this angular distribution. The vector polarization (tensor of rank 1) behaves like a classical vector and is fully described by its three components. It points into a distinct *direction* which is characterized by a sign—different from the tensor polarization which has only an orientation ("*alignment*") of an anisotropic spin distribution with respect to an *axis*.

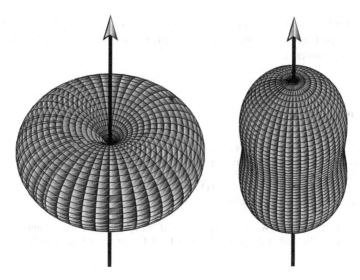

Fig. 3.6 Graphical representation of the spin (tensor) anisotropy of a spin-1 ensemble polarized completely in the z direction with negative tensor polarization (corresponding to relative occupation numbers of $N_0 = 1$, $N_+ = N_- = 0$) (*left*) and with positive tensor polarization (corresponding to relative occupation numbers of either $N_+ = 1$, $N_0 = N_- = 0$ or $N_- = 1$, $N_0 = N_+ = 0$) (*right*)

3.6.9 Polarization of Particles with Higher Spin

The fact that particles with $S = 3/2$ (^{7}Li, ^{23}Na) have been polarized (see Sect. 8.4.2) requires an appropriate description of tensor moments. This case has been discussed in Ref. [6], so will not be detailed here. It is clear that tensor moments up to rank three have to be considered. They must be constructed from spin operators for $S = 3/2$ which—like in the spin-1/2 and spin-3/2 case (see Eqs. 3.57, 3.95, and 3.86)—can be derived from contractions of lower-spin operators and taking into account commutation relations and transformation properties of spherical tensors as well as a normalization condition. Note that in the following equations spherical tensor moments are thus expressed by spherical combinations of the spin-1 Cartesian tensors. :

$$t_{30} = \frac{1}{6\sqrt{5}} \left\langle 20S_z^3 - 41S_z \right\rangle$$

$$t_{3\pm1} = \mp \frac{1}{24\sqrt{15}} \left\langle \left(60S_z^2 - 51\right)(S_x \pm iS_y) + (S_x \pm iS_y)\left(60S_z^2 - 51\right) \right\rangle$$

$$t_{3\pm2} = \frac{1}{\sqrt{6}} \left\langle S_z (S_x \pm iS_y)^2 + (S_x \pm iS_y)^2 S_z \right\rangle$$

$$t_{3\pm3} = \mp \frac{1}{3} \left\langle (S_x \pm iS_y)^3 \right\rangle \tag{3.107}$$

Though spins > 1 have not been considered in the Madison convention its application, e.g. concerning coordinate systems is straightforward. It is evident that for higher spins the Cartesian notation becomes impractical and spherical tensor moments should be used (the term "efficiency tensor (moment)" in [6] should be replaced by "analyzing power").

References

1. Fano, U.: Rev. Mod. Phys. **29**, 74 (1957)
2. Feynman, R.P.: Statistical Mechanics, pp. 39. Benjamin, Reading, MA (1972)
3. Fano, U.: Phys. Rev. **90**, 577 (1953)
4. Ohlsen, G.G.: Rep. Progr. Phys. **35**, 717 (1972)
5. Lakin, W.: Phys. Rev. **98**, 139 (1958)
6. Darden, S.E.: In: [7] p. 39 (1971)
7. Barschall, H.H., Haeberli, W. (eds.): Proceedings of 3rd International Symposium on Polarization Phenomena in Nuclear Reactions. Madison 1970, University of Wisconsin Press (1971)

Chapter 4
Rotations, Angular Dependence of the Tensor Moments

4.1 Generalities

It is important to be able to describe polarization observables in rotated coordinate systems. Typical applications are: Spin precession in magnetic fields, deflection of polarized particle beams by optical elements, nuclear reactions, double scattering and polarization transfer. We start with the rotation of the density matrix ρ.

A reminder:

In quantum mechanics finite rotations of a system are described by rotation operators which are integrals over operators for infinitesimal rotations (which are linear!). An example: the rotation operator for the finite rotation by an angle α around the z axis has the form (the components of S are given in units of $\hbar$ (i.e. $\hbar \equiv 1$)):

$$\mathbf{D}(\alpha) = e^{-i\alpha S_z} \tag{4.1}$$

4.2 The Description of Rotations by Rotation Operators

The most general rotation is composed from three rotations by the Euler angles (Attention! Several conventions exist; her we follow Condon/Shortley [1], Rose [2] and Brink/Satchler [3]): a sequence of right-handed rotations about the z, then the new y, and then about the then new z axis. This is equivalent to a sequence of rotations about the respective old z, y, and x axes. In addition, one has to distinguish between "active" rotations of the system in a fixed coordinate system and "passive" rotations of the coordinate system. A very comprehensive and useful discussion of conventions and different definitions of rotation functions, angular-momentum coupling coefficients, irreducibility etc. can be found in [4], especially in Tables 3.1, 4.1, 4.2, and 5.1. The definition of rotation operators used here is:

$$\mathbf{D}(\alpha\beta\gamma) = e^{-i\gamma S_z'} e^{-i\beta S_y''} e^{-i\alpha S_z}$$
$$= e^{-i\alpha S_z'} e^{-i\beta S_y} e^{-i\gamma S_z} = e^{-i\beta S_y} e^{-i(\alpha+\gamma)S_z} \tag{4.2}$$

H. Paetz gen. Schieck, *Nuclear Physics with Polarized Particles*,
Lecture Notes in Physics 842, DOI: 10.1007/978-3-642-24226-7_4,
© Springer-Verlag Berlin Heidelberg 2012

They have the matrix elements:

$$\langle IM'|\mathbf{D}(\alpha\beta\gamma)|IM\rangle = D^I_{MM'}(\alpha\beta\gamma) \tag{4.3}$$

and

$$\langle IM'|\mathbf{D}^\dagger(\alpha\beta\gamma)|IM\rangle = D^I_{MM'}{}^*(\alpha\beta\gamma) \tag{4.4}$$

D is unitary, i.e.

$$\mathbf{D}^\dagger(\alpha\beta\gamma) = \mathbf{D}^{-1}(\alpha\beta\gamma) = \mathbf{D}(-\alpha, -\beta, -\gamma) \tag{4.5}$$

i.e.

$$D^I_{MM'}{}^*(\alpha\beta\gamma) = D^I_{M'M}(-\gamma, -\beta, -\alpha) \tag{4.6}$$

and

$$\sum_{M'} D^I_{M'N}{}^* D^I_{M'M} = \delta_{MN}. \tag{4.7}$$

For the product (two rotations in sequence):

$$D^L_{MM'} = \sum_{m_1 m_1' m_2 m_2'} D^{j_1}_{m_1 m_1'} D^{j_2}_{m_2 m_2'} (j_1 j_2 m_1 m_2 | LM)(j_1 j_2 m_1' m_2' | LM') \tag{4.8}$$

(The symbol $(j_1 j_2 m_1 m_2 | LM)$ denotes the Clebsch-Gordan (vector coupling) coefficients.)

When choosing as usual the eigenfunctions of S_z as basis vectors **D** simplifies to:

$$D^I_{MN}(\alpha\beta\gamma) = e^{-i(\alpha M + \gamma N)}\langle IM|e^{-i\beta S_y}|IN\rangle \tag{4.9}$$

The first factor is a phase factor, the second a matrix element of the *reduced rotation functions* $d^I_{MN}(\beta)$. Condon/Shortley [1] define the d^k_{mn} as:

$$d^k_{mn} = \sum_t (-)^t \frac{[(k+m)!(k-m)!(k+n)!(k-n)!]^{1/2}}{(k+m-t)!(k-n-t)!t!(t+n-m)!}$$

$$\times \left(\cos\frac{\beta}{2}\right)^{2k+m-n-2t} \left(\sin\frac{\beta}{2}\right)^{2t+n-m} \tag{4.10}$$

The sum goes over all t leading to non-negative factorials. For M or $N = 0$ **D** becomes a spherical harmonic via:

$$D^I_{M0}(\beta\alpha) = \left(\frac{4\pi}{2I+1}\right)^{1/2} Y^M_I(\beta, \alpha) = \left(\frac{4\pi}{2I+1}\right)^{1/2} Y^{-M}_I(\beta, \alpha) \tag{4.11}$$

and:

$$D_{00}^I(\beta) = P_I(\cos\beta) \quad \text{(Legendre polynomial)} \tag{4.12}$$

Example $S = 1/2$, $S_y = 1/2\sigma_y$:

$$e^{-i\beta S_y} = e^{-i\beta\sigma_y/2} = \cos\frac{\beta}{2} - i\sigma_y\sin\frac{\beta}{2} \tag{4.13}$$

From here the *reduced rotation function* results for a rotation about the y axis:

$$d_{mm'}^{1/2} = \begin{pmatrix} \cos\frac{\beta}{2} & -\sin\frac{\beta}{2} \\ \sin\frac{\beta}{2} & \cos\frac{\beta}{2} \end{pmatrix} \tag{4.14}$$

4.3 Rotation of the Density Matrix and of the Tensor Moments

This allows e.g. a description of the density matrix in a rotated system or—equivalently—of a rotated density matrix in the old coordinate system. As an example we rotate the density matrix which is diagonal with respect to the z axis:

$$\begin{aligned} \rho' &= d_{mm'}^{1/2}(\beta)\frac{1}{2}\begin{pmatrix} 1+p & 0 \\ 0 & 1-p \end{pmatrix} d_{mm'}^{1/2\,\dagger}(\beta) \\ &= d_{mm'}^{1/2}(\beta)\frac{1}{2}\begin{pmatrix} 1+p & 0 \\ 0 & 1-p \end{pmatrix} d_{mm'}^{1/2\,T}(\beta) \end{aligned} \tag{4.15}$$

The polarization direction is completely determined by the two parameters β, ϕ. β is the polar angle (relative to the z axis), ϕ the azimuthal angle = angle of the $(\vec{S}, z)$ plane relative to an arbitrarily defined x axis (where x, y, z form a righthanded system). In the rotation function **D** therefore only two angular parameters are physically relevant:

$$D_{MM'}^I(\alpha, -\tilde{\beta}_{\text{Euler}}, -\gamma - \pi/2) \equiv D_{MM'}^I(0, \beta_{\text{polar}}, \phi_{\text{azimut}}) \tag{4.16}$$

Under rotations tensor moments transform more simply than the density matrix. The spherical tensors have been defined such as to transform under rotations like the spherical harmonics:

$$t_{kq} = \sum_{q'} D_{q'q}^k(\alpha\beta\tilde{\gamma})t_{kq'} = \hat{t}_{k0}D_{0q}^k(\beta, \phi) \tag{4.17}$$

Here $\hat{t}_{k0}$ signifies the maximum component for which ρ also is diagonal, i.e. if the z axis is the quantization axis. Thus the single components for spin S = 1 (S = 1/2 is trivial) are:

$$t_{10} = \hat{t}_{10} \cos \beta$$

$$t_{1\pm 1} = \mp i \hat{t}_{10} \frac{1}{\sqrt{2}} \sin \beta e^{i\phi}$$

$$t_{20} = \hat{t}_{20} \frac{1}{2}(3 \cos^2 \beta - 1) \tag{4.18}$$

$$t_{21} = -i \hat{t}_{20} \sqrt{\frac{3}{2}} \sin \beta \cos \beta e^{i\phi}$$

$$t_{22} = -\hat{t}_{20} \sqrt{\frac{3}{8}} \sin^2 \beta e^{2i\phi}$$

For the Cartesian tensors the corresponding transformations result from the connection with the spherical tensors. Example for p_{zz}: $p_{zz} = p_{zz}^* P_2(\cos \beta)$, where p^* and p_{zz}^* are the *Cartesian* maximum components of the vector and tensor polarization for S in the direction z. Thus:

$$p_x = p^* \cdot P_1^1(\cos \beta) \cos \phi = p^* \cdot \sin \beta \cos \phi$$

$$p_y = p^* \cdot P_1^1(\cos \beta) \sin \phi = p^* \cdot \sin \beta \sin \phi$$

$$p_z = p^* \cdot P_1(\cos \beta) = p^* \cdot \cos \beta$$

$$p_{zz} = p_{zz}^* \cdot P_2(\cos \beta) = p_{zz}^* \cdot \frac{1}{2}(3 \cos^2 \beta - 1)$$

$$p_{xx} = p_{zz}^* \cdot \frac{1}{2}\left[P_2^2(\cos \beta) \cos^2 \phi - 1 \right]$$

$$p_{yy} = p_{zz}^* \cdot \frac{1}{2}\left[P_2^2(\cos \beta) \sin^2 \phi - 1 \right]$$

$$p_{xx} - p_{yy} = p_{zz}^* \cdot \frac{1}{2} P_2^2(\cos \beta) \cos 2\phi \tag{4.19}$$

$$= p_{zz}^* \cdot \left(-\frac{3}{2}\right) \sin 2\beta \cos 2\phi$$

$$p_{xy} = p_{zz}^* \cdot P_2^2(\cos \beta) \sin 2\phi$$

$$p_{yz} = p_{zz}^* \cdot \frac{1}{2} P_2^1(\cos \beta) \sin \phi$$

$$p_{xz} = p_{zz}^* \cdot \frac{1}{2} P_2^1(\cos \beta) \cos \phi = p_{zz}^* \cdot \left(-\frac{3}{4}\right) \sin 2\beta \cos \phi$$

Experimentally interesting special cases are:

- $[\beta = 0]°$:   only p_z, $p_{zz} \neq 0$
- $[\beta = 54.7]°$: $p_{zz} = 0$
- $[\beta = 90]°$ and e.g. $[\phi = 90]°$:

$$p_{zz} = -\frac{1}{2}p_{zz}^*$$
$$p_{xx} - p_{yy} = \frac{3}{2}p_{zz}^*$$
$$p_y = p^*$$
$$\text{and} \quad p_x = p_z = p_{xz}$$
$$= p_{yz} = p_{xy} = 0.$$

4.4 Practical Realization of Rotations

In practice the rotation of the quantization axis is achieved using the Larmor precession of the spins in suitable magnetic fields. Only components perpendicular to the
spin vector are affected. For charged particles the deflection of a beam is coupled
to the spin precession (e.g. in dipole and quadrupole magnets etc.) The component
parallel to the magnetic field remains unchanged. In Wien filters the magnetic deflection is compensated by an electric field perpendicular to the magnetic field and the
velocity vector of the particles. For homogeneous fields the Wien filter is "straight-
looking", when (in MKSA units) the velocity of the particles is v = E/B. With a
Wien filter rotatable about the beam axis any spin direction in space can be realized
(see Sect. 8.6).

4.5 Coordinate System

In nuclear reactions the outgoing beam in general will be rotated against the incident
beam by the scattering angles Θ, Φ. The tensor moments in the entrance and exit
channels can be described in different ways. Very common is the description in the
helicitiy formalism in which each particle is described with respect to its direction
of motion $\vec{k}_{in}$ and $\vec{k}_{out}$, respectively, as its positive z axis. The positive y axis follows
the convention

$$\hat{y} = \frac{\vec{k}_{in} \times \vec{k}_{out}}{|\vec{k}_{in} \times \vec{k}_{out}|} \tag{4.20}$$

while the x axis is determined by requiring a right-handed system. In going from the
entrance to the exit channel there is the choice of describing the rotation about angles
defined either in the laboratory or in the relative (c.m.) system. This convention is
the *Madison convention* [5] and implies the *Basel convention* [6], see Sect. 5.4.

References

1. Condon, E.U., Shortley, G.H.: The Theory of Atomic Spectra. Cambridge University Press, Cambridge (1967)
2. Rose, M.E.: Elementary Theory of Angular Momentum. Wiley, New York (1957)
3. Brink, D.M., Satchler, G.R.: Angular Momentum. Oxford University Press, Oxford (1971)
4. Chaichian, M., Hagedorn, R.: Symmetries in Quantum Mechanics—From Angular Momentum to Supersymmetry, Graduate Student Series in Physics. Institute of Physics Publishing, Bristol (1998)
5. Barschall, H.H., Haeberli, W. (eds.): Proceedings of the 3rd International Symposium on Polarization Phenomena in Nuclear Reactions, Madison 1970. University of Wisconsin Press, Madison (1971)
6. Huber, P., Meyer, K.P. (eds.): Proceedings of the International Symposium on Polarization Phenomena of Nucleons, Basel 1960. Helv. Phys. Acta Suppl. VI. Birkhäuser, Basel (1961)

Part II
Nuclear Reactions

Chapter 5
Description of Nuclear Reactions of Particles with Spin

5.1 General

As for nuclear reactions of spinless particles (this is the only case and the simplest one normally treated in lectures and textbooks) scattering amplitudes between entrance and exit states are a useful tool for the description of two-particle nuclear reactions, $a + A \longrightarrow b + B$ (or $A(a, b)B$). Now these states are, however, spin substates in the entrance and exit channels which leads to the following complications:

- A scattering amplitude will become a scattering matrix.
- Depending on which quantum numbers are conserved (which depends on the symmetry properties of the physical system) angular-momentum coupling has to be performed, e.g. the spins of the incident particles s_a, s_A are coupled to the entrance channel spin S, this in turn is coupled to the entrance-channel orbital angular momentum l to yield the (conserved) total angular momentum J, and analogously for the exit channel (’):

$$S(s_a + s_A) + l \longrightarrow J \longrightarrow S'(s_b + s_B) + l' \tag{5.1}$$

It is useful to apply the formalism of Racah algebra (6j, 9j symbols, or Wigner's W or Z coefficients) in which the summations about magnetic quantum numbers [1] have been performed already and which have relatively simple symmetries and rules.

- In two-particle reactions the entrance and the exit channels may contain up to two particles with spin. The description of the spin state of the entrance and exit channel takes place in a spin state the dimension of which is the direct product of the spin-space dimensions of each of the two particles: $(2s_a + 1)(2s_A + 1)$ and $(2s_b + 1)(2s_B + 1)$. The corresponding density matrices are also products of the two sub-density matrices and similarly for their expansions into Cartesian or spherical spin tensors. Since in the entrance channel normally there exist no correlations between the spin states of the beam and target the density matrix for the entrance

H. Paetz gen. Schieck, *Nuclear Physics with Polarized Particles*,
Lecture Notes in Physics 842, DOI: 10.1007/978-3-642-24226-7_5,
© Springer-Verlag Berlin Heidelberg 2012

channel can be factorized as well as the corresponding spin tensor. This, however, never applies for the exit channel.

5.2 The M Matrix

The generalization of the scattering ampitude is the transfer or M matrix. It is the matrix which transforms the entrance channel density matrix into that of the exit channel . Formally:

$$\rho_{fin} = \mathbf{M}\rho_{in}\mathbf{M}^\dagger \tag{5.2}$$

ρ_{in} and ρ_{fin} are the direct-product density matrices of the two particles of the entrance and the exit channel, respectively:

$$\rho_{in} = \rho_a(s_a) \otimes \rho_A(s_A) \quad \text{and} \quad \rho_{fin} = \rho_b(s_b) \otimes \rho_B(s_B) \tag{5.3}$$

This allows in principle a complete description of a nuclear reaction, more precisely of its observables which have been defined as expectation values of certain (spin) operators.

(a) Besides the inegrated (total) cross-section the simplest observable is the unpolarized differential cross-section, for which the beam and target are unpolarized and no polarization, but only intensities are measured in the exit channel. It is defined as being proportional to the normalized expectation value of the unit operator ("intensity"):

$$W = \text{Tr}(\rho_{fin}E) = \text{Tr}(M\rho_{in}M^\dagger E) \tag{5.4}$$

With

$$\rho_{in} = \rho_a(s_a) \otimes \rho_A(s_A) = \frac{1}{2s_a + 1}E(s_a) \times \frac{1}{2s_A + 1}E(s_A) \tag{5.5}$$

[this is a $(2s_a + 1)(2s_A + 1) \times (2s_a + 1)(2s_A + 1)$ matrix] follows:

$$\rho_{fin} = \frac{1}{(2s_a + 1)(2s_A + 1)}\mathbf{M}\mathbf{M}^\dagger \tag{5.6}$$

This results in the cross-section in the usual sense—the "unpolarized" cross-section, if one appplies its "usual" definition ("outgoing particle current into the solid-angle element $d\Omega$ at the angle Θ, divided by the incident particle current density") and the correct phase-space factors (density of final states, Fermi's "golden rule"):

$$(d\sigma/d\Omega)_0 = \frac{1}{(2s_a + 1)(2s_A + 1)}\frac{k_{fin}}{k_{in}}\text{Tr}(\mathbf{M}\mathbf{M}^\dagger) \tag{5.7}$$

(b) When the incident beam is polarized, the target unpolarized, no polarization in the exit channel is measured and an expansion of the density matrix into Cartesian or spherical tensors with their special transformation properties is used, then ρ_{in} may be written e.g. in Cartesian coordinates:

$$\rho_{in} = \frac{1}{(2s_a + 1)(2s_A + 1)} \sum_i \overbrace{\langle \sigma_i \rangle}^{p_i} (\sigma_i \otimes E(s_A)) \tag{5.8}$$

and the cross-section:

$$\begin{aligned}
d\sigma/d\Omega &= \frac{1}{(2s_a + 1)(2s_A + 1)} \frac{\mathbf{k}_{fin}}{\mathbf{k}_{in}} \sum_i p_i \, \mathrm{Tr}(\mathbf{M}\sigma_i \mathbf{M}^\dagger) \\
&= \frac{1}{(2s_a + 1)(2s_A + 1)} \frac{\mathbf{k}_{fin}}{\mathbf{k}_{in}} \left[\mathrm{Tr}(\mathbf{M}\mathbf{M}^\dagger) + \sum_i p_i \, \mathrm{Tr}(\mathbf{M}\sigma_i \mathbf{M}^\dagger) \right] \\
&= (d\sigma/d\Omega)_0 \left[1 + \sum_i p_i \frac{\mathrm{Tr}(\mathbf{M}\sigma_i \mathbf{M}^\dagger)}{\mathrm{Tr}(\mathbf{M}\mathbf{M}^\dagger)} \right]
\end{aligned} \tag{5.9}$$

The $A_i = \frac{\mathrm{Tr}(\mathbf{M}\sigma_i \mathbf{M}^\dagger)}{\mathrm{Tr}(\mathbf{M}\mathbf{M}^\dagger)}$ are *analyzing powers* which measure how the reaction is influenced by each single component of the beam polarization p_i. The index i has a general meaning: for $s_a = 1/2$ the beam can at most be vector polarized (rank 1) and the range of i is 1–3 or x, y, z, respectively. For spin-1 particles i signifies the components of the vector polarization and of the rank-2 tensor polarization [in the Cartesian case i signifies the combinations x, y, z and (jk) with j, k = x, y, z]. Therefore these are the components of the vector and tensor analyzing powers.

(c) When the polarization of an outgoing particle (e.g. the ejectile b) is measured while the incident beam is unpolarized it is defined as

$$\begin{aligned}
\vec{p}' &= \frac{\mathrm{Tr}(\vec{\sigma} \rho_{fin})}{\mathrm{Tr}(\rho_{fin})} = \frac{\mathrm{Tr}(\vec{\sigma} \mathbf{M} \rho_{in} \mathbf{M}^\dagger)}{\mathrm{Tr}(\rho_{fin})} \\
&= \frac{(2s_a + 1)(2s_b + 1)}{(2s_A + 1)(2s_B + 1)} \frac{\mathrm{Tr}(\mathbf{M}\mathbf{M}^\dagger \vec{\sigma})}{\mathrm{Tr}(\mathbf{M}\mathbf{M}^\dagger)}
\end{aligned} \tag{5.10}$$

where the transformation properties of $\vec{p}'$ are those of the components of $\vec{\sigma}$ for the spin system considered. If, e.g. the $\vec{\sigma}$ are the spin-1/2 Pauli operators $\vec{p}$ will be a vector with the Cartesian components p_x, p_y, and p_z.

(d) More complicated cases are those where e.g. both particles in the entrance channel are polarized or the polarizations of both exit-channel particles (ejectile and residual nucleus) are measured in coincidence ("spin correlations") or those in which the influence of the polarization(s) in the entrance channel on the polarization(s) in the exit channel is measured ("polarization transfer coefficients", "triple-scattering parameters"). The term "triple scattering" stems from the beginning of polarization experiments when for the production of polarized beams a primary nuclear reaction was needed, the second "scattering" was the one to be investigated, and the third reaction served as an analyzer for the polarization of the outgoing particles (see Chap. 12). With the advent of polarized-ion sources only "double scattering" was required. For a consistent description especially of more complicated spin states the spherical one is best suited. The two-particle spin tensors are defined as:

- for the entrance channel: $\tau_{kqKQ} = \tau_{kq}\tau_{KQ}$, since in this case polarizations of projectile and target are uncorrelated
- for the exit channel: $\tau_{k'q'K'Q'}$.

Lower-case indices are for projectile or ejectile, resp., upper-case ones for the target or the recoil nucleus. The most general polarization observable therefore depends on the polarization state (or on this state being measured!) of maximally four particles, i.e. in spherical notation on eight indices:

$$\tau_{kqKQ}^{k'q'K'Q'} \tag{5.11}$$

5.3 Types of Polarization Observables

Besides the most general notation [i.e. the spherical one with kq for the projectile, KQ for the target, k'q' for the ejectile and K'Q' for the recoil (residual) nucleus] the simplified Saclay description in the helicity coordinate system is being used: X_{pqik} with X defining the observable (e.g. = A for analyzing power, C for correlations etc.) [2]. The pqik designate the ejectile, the recoil nucleus, the projectile (the beam) and the target, respectively; their values are: s ("sideways"), n ("normal"), l ("longitudinal") for polarized particles and 0 for unpolarized ones. Thus the following types of observables can be classified (besides the character and name of the observable the nomenclature for the reaction, a typical Cartesian example and the spherical definition is listed):

- Zero-spin observable: Unpolarized cross-section X_{0000} or I_{0000}:

$$A(a,b)B \qquad (d\sigma/d\Omega)_0 \propto \mathrm{Tr}(\mathbf{M}\mathbf{M}^\dagger) \tag{5.12}$$

- One-spin observables

  - Projectile analyzing power, e.g. A_{00i0} or A_y:

$$A(\vec{a},b)B \qquad T_{kq} \propto \frac{\mathrm{Tr}(\mathbf{M}\tau_{kq}\mathbf{M}^\dagger)}{\mathrm{Tr}(\mathbf{M}\mathbf{M}^\dagger)} \tag{5.13}$$

  - Target analyzing power, e.g. A_{000k} or A_y:

$$\vec{A}(a,b)B \qquad T_{KQ} \propto \frac{\mathrm{Tr}(\mathbf{M}\tau_{KQ}\mathbf{M}^{\dagger})}{\mathrm{Tr}(\mathbf{M}\mathbf{M}^{\dagger})} \tag{5.14}$$

- Outgoing polarization, e.g. p_{p000} or $p_{y'}$:

$$A(a,\vec{b})B \qquad t_{k'q'} \propto \frac{\mathrm{Tr}(\mathbf{M}\mathbf{M}^{\dagger}\tau_{k'q'})}{\mathrm{Tr}(\mathbf{M}\mathbf{M}^{\dagger})} \tag{5.15}$$

- Similarly for the recoil nuclei, e.g. p_{0q00}:

$$A(a,b)\vec{B} \qquad t_{K'Q'} \propto \frac{\mathrm{Tr}(\mathbf{M}\mathbf{M}^{\dagger}\tau_{K'Q'})}{\mathrm{Tr}(\mathbf{M}\mathbf{M}^{\dagger})} \tag{5.16}$$

- *Two-spin observables*

  - Polarization transfer coefficients (In the case of the NN interaction they were also designated as "Wolfenstein parameters" $A(= K_z^{x'})$, $A'(= K_z^{z'})$, $R(= K_x^{x'})$, $R'(= K_x^{z'})$, $D(= K_y^{y'})$). Notation: D_{p0i0}, e.g. Cartesian $K_y^{y'}$, or spherical: e.g. $t_{kq}^{k'q'}$. Meaning: transfer of the projectile polarization to that of the ejectile:

$$A(\vec{a},\vec{b})B \qquad t_{kq}^{k'q'} \propto \frac{\mathrm{Tr}(\tau_{k'q'}\cdot\mathbf{M}\tau_{kq}\mathbf{M}^{\dagger})}{\mathrm{Tr}(\mathbf{M}\mathbf{M}^{\dagger})} \tag{5.17}$$

  - Spin correlation coefficients: Correlation either in the entrance or the exit channel (in the first case the polarizations of beam and target are of course uncorrelated, in the latter case the are correlated by the nuclear reaction and therefore must be measured in coincidence). The spin tensors are direct products of the tensors of both particles and act e.g. for the entrance channel in a $(2s_a + 1)(2s_A + 1)$ dimensional spin space. Notation: In the case of the NN interaction e.g. A_{00nn}, Cartesian A_{mn} with $m, n = x, y, z$ (not to be mixed up with the tensor analyzing power A_{ik}) means an entrance-channel correlation of the beam and target polarizations in y (= normal) direction, or $C_{ik,\ell m}$; more generally the spherical notation $t_{kqKQ} = t_{kqKQ}^{0000}$ and $t^{k'q'K'Q'} = t_{0000}^{k'q'K'Q'}$, respectively, may be used. For entrance-channel correlations:

$$\vec{A}(\vec{a},b)B \qquad t_{kq}t_{KQ} \propto \frac{\mathrm{Tr}(\mathbf{M}\tau_{kqKQ}\mathbf{M}^{\dagger})}{\mathrm{Tr}(\mathbf{M}\mathbf{M}^{\dagger})} \tag{5.18}$$

with factorized input tensor moments and similarly (with correlated outgoing tensor moments) for the exit channel.

- *Three- and four-spin observables*: Analogous expressions hold for "generalized analyzing powers" $t_{qkQK}^{q'k'Q'K'}$ using more general spherical (or Cartesian) tensors $\tau_{kqKQ}^{k'q'K'Q'}$.

5.4 Coordinate Systems

The notation of polarization tensors (tensor moments) and suitable coordinate systems have been strongly recommended in two international conventions:

- The *Basel Convention* was issued in 1960 [3] and determines that in nuclear reactions with spin-1/2 particles the polarization should be counted positive in the direction $\vec{k}_{in} \times \vec{k}_{out}$. Assuming a positive analyzing power this positive polarization yields a positive left–right asymmetry (L–R).
- The *Madison Convention* [4, 5] refers to spin-1 particles. A right-handed coordinate system is assumed with the z direction being the momentum direction of the incident or outgoing particles, and where the y direction is along $\vec{k}_{in} \times \vec{k}_{out}$. Cartesian and spherical spin tensors and the corresponding tensor moments are allowed and the components of the polarization are given by p_i, p_{ij} (Cartesian) or t_{kq} (spherical), respectively, those of the analyzing powers A_i, A_{ij}; $i, j = x, y, z$ (Cartesian) or T_{kq}, respectively. The extension to higher spins is straightforward leading to an increasing number of indices for the Cartesian notation.

One consequence of this convention is that when using more than one detector each detector obtains its own coordinate system with $\hat{y}$ perpendicular to the respective scattering plane. Figure 5.1 shows the situation for one system with polarized particles.

5.4.1 Coordinate Systems for Analyzing Powers

Although the (spin) observables to be measured depend only on the polar angle Θ, the cross-sections including these observables generally exhibit a dependence on the azimuthal angle ϕ. This dependence enters via the need to introduce coordinate systems in which the detector positions as well as the polarization direction have to be described. While for the vector polarization the ensuing *azimuthal complexity* is a simple $\sin\phi$ or $\cos\phi$ dependence, e.g. for spin 1/2 with the Madison convention and parity conservation

$$
\begin{aligned}
\frac{d\sigma}{d\Omega}(\Theta, \phi) &= \left(\frac{d\sigma}{d\Omega}\right)_0 (\Theta)\left[1 + \frac{1}{2}p_y A_y(\Theta)\right] \\
&= \left(\frac{d\sigma}{d\Omega}\right)_0 \left[1 + \frac{1}{2}p_z A_y(\Theta)\sin\beta\cos\phi\right],
\end{aligned}
\tag{5.19}
$$

for higher spins and other types of observables the description is more complex. As an example the case of polarized spin-1 projectiles such as deuterons on unpolarized targets will be discussed. Writing out the expansion Eq. 5.9 in Cartesian notation and considering parity conservation, as discussed below, we obtain (the cross-sections and analyzing powers depend only on Θ and the energy)

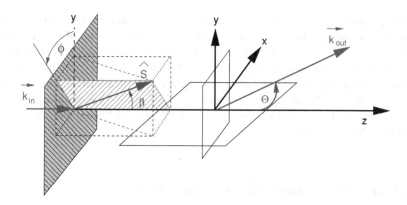

Fig. 5.1 Coordinate system for the description of the incident polarization as well as construction of the y axis with x along $\vec{y} \times \vec{k}_{in}$, y along $\vec{k}_{in} \times \vec{k}_{out}$, and z along $\vec{k}_{in}$

$$\frac{d\sigma}{d\Omega} = \left(\frac{d\sigma}{d\Omega}\right)_0 \left[1 + \frac{3}{2}p_y A_y + \frac{1}{2}p_{zz}A_{zz} + \frac{2}{3}p_{xz}A_{xz} + \frac{1}{6}(p_{xx} - p_{yy})(A_{xx} - A_{yy})\right]$$

$$(5.20)$$

After introducing the β and ϕ dependences of the quantization (spin-symmetry) axis (of the Madison convention [6], see also Fig. 5.1) and parity conservation explicitly we obtain

$$\frac{d\sigma}{d\Omega} = \left(\frac{d\sigma}{d\Omega}\right)_0 \left\{ 1 + \frac{3}{2}p_Z A_y \sin\beta \cos\phi \right.$$

$$+ p_{ZZ}\left[\frac{1}{4}A_{zz}(3\cos^2\beta - 1) - A_{xz}\sin\beta\cos\beta\sin\phi\right.$$

$$\left.\left. - \frac{1}{4}(A_{xx} - A_{yy})\sin^2\beta\cos 2\phi\right]\right\},$$

$$(5.21)$$

where p_Z and p_{ZZ} (often: p_z^* or p^* and p_{zz}^*, also $\hat{p}_Z$ and $\hat{p}_{ZZ}$, see also Eq. 5.22 for spin correlations) are the coordinate-independent maximum values along the quantization (symmetry) axis of the polarization, e.g. of a beam coming from a polarized-ion source or of a polarized target. We see that the maximum azimuthal complexity is $\propto \cos 2\phi$ which has to be taken into account for the placement of detectors. An arrangement of four detectors $\Delta\phi = [90]°$ apart at one polar angle Θ is advantageous, because by taking differences and sums of count ratios of the four detectors each of the four analyzing powers can be determined nearly independently of all others [7].

5.4.2 Coordinate Systems for Polarization Transfer

For the determination of polarization-transfer coefficients the polarization of the ejectiles from a primary reaction, induced with polarized particles, has to be measured ("double scattering"). This is done using a (calibrated) analyzer reaction. For this again the Madison convention is used, i.e. the direction of motion of the outgoing particles (along $\vec{k}_{out}$ in Fig. 5.1) is the new z' axis for the second scattering. However, this axis can be defined along the c.m. or the lab. direction. Then the coordinate system for the analyzer reaction may be defined as before as a right-handed system with: x' along $\vec{y}' \times \vec{z}'$, y' along $\vec{z} \times \vec{k}'_{out}$, and $z' = \vec{k}'_{out}$. More detailed discussions on polarization transfer can be found in Ohlsen [8], Ohlsen and Keaton [9], Sperisen et al. [10], Sydow et al. [11, 12].

5.4.3 Coordinate Systems for Spin Correlations

In this case two—in principle independent—polarizations have to be considered, because they can be prepared independently with arbitrary spin directions. With the above prescription for both, two coordinate systems can be defined which will be rotated azimuthally against each other around the z axis, i.e. they are connected by one azimuthal angle Φ. The azimuthal angles describing the detector positions in both systems are therefore connected by trigonometric relations containing Φ.

As mentioned above the observables (such as spin-correlation coefficients) depend only on the polar angle Θ whereas the (polarized) cross section generally acquires an azimuthal dependence via the introduction of coordinate systems. These are in principle arbitrary but we follow Ohlsen [8] and the Madison Convention [4]. In Ohlsen [8] the case of the azimuthal dependence of spin-correlation cross sections is explained for spin-1/2 on spin-1 systems (see also [13]). For the spin-1 on spin-1 case this is explained in Paetz gen. Schieck [14].

Genrally, it is advisable to choose such a system that the description of a real experiment is as simple and intuitive as possible. In this respect the Cartesian description is more intuitive than the spherical one. The description of polarization components from polarized sources and polarized targets is best imagined in a space-fixed coordinate system in which the direction of the polarization vectors are described by two sets of polar and azimuthal angles (β_b, ϕ_b) for the incident beam polarization and (β_t, ϕ_t) for the target. The orientation of the tensor polarization is fixed to that of the polarization vector. As coordinate system here a set of axes x, y, and z is chosen, where z is identical with the incoming beam direction (along $\vec{k}_{in}$), y may be vertically upward, and x, y, and z form a right-handed screw.

On the other hand, we need a scattering-frame system where a Y axis is defined by the direction of $\vec{k}_{in} \times \vec{k}_{out}$, the Z axis coincides with z, and with the X-axis again forming a right-handed system together with $Z = z$. It is clear that this system is different for each detector, the position of which must be characterized by a polar angle Θ and some azimuthal angle. We demand that for the parts of the cross section with only one particle type (beam or target) being polarized (leading to analyzing powers) we have the usual description of the azimuthal dependence on ϕ (with a maximum azimuthal complexity of $\cos 2\phi$), see Sect. 5.4.1. Figure 5.2 shows the

Fig. 5.2 Coordinate systems for spin-1 on spin-1 polarization correlation experiments. The polarization symmetry axes S_b for beam (b) and S_t for target (t) polarizations are defined in the space-fixed coordinate system x, y, *and* z. The detector(s) are positioned at polar angles Θ with respect to the $z = Z$ axis and at angles ϕ as measured clockwise from the x axis along z. Relative to the spin directions the azimuthal angles are $\phi_b - \Phi$ and $\phi_t - \Phi$. The polar angles of the polarizations β_b and β_t are not shown

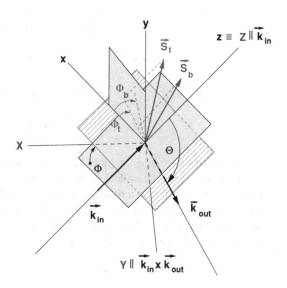

relations between the two polarization-symmetry axes and the projectile-helicity frame. The polarization components in the scattering frame of the incident beam are

$$p_X = \hat{p}_Z \sin \beta_b \cos(\phi_b - \Phi)$$
$$p_Y = \hat{p}_Z \sin \beta_b \sin(\phi_b - \Phi)$$
$$p_Z = \hat{p}_Z \cos \beta_b$$
$$p_{XY} = \frac{3}{4} \hat{p}_{ZZ} \sin^2 \beta_b \sin 2(\phi_b - \Phi)$$
$$p_{YZ} = \frac{3}{2} \hat{p}_{ZZ} \sin \beta_b \cos \beta_b \sin 2(\phi_b - \Phi)$$
$$p_{XZ} = \frac{3}{2} \hat{p}_{ZZ} \sin \beta_b \cos \beta_b \cos(\phi_b - \Phi)$$
$$p_{XX} - p_{YY} = \frac{3}{2} \hat{p}_{ZZ} \sin^2 \beta_b \cos 2(\phi_b - \Phi)$$
$$p_{ZZ} = \frac{1}{2} \hat{p}_{ZZ}(3 \cos^2 \beta_b - 1)$$

(5.22)

and similarly for the target polarization

$$q_X = \hat{q}_Z \sin \beta_t \cos(\phi_t - \Phi)$$
$$q_Y = \hat{q}_Z \sin \beta_t \sin(\phi_t - \Phi)$$
$$q_Z = \hat{q}_Z \cos \beta_t$$
$$q_{XY} = \frac{3}{4} \hat{q}_{ZZ} \sin^2 \beta_t \sin 2(\phi_t - \Phi)$$

$$q_{YZ} = \frac{3}{2}\hat{q}_{ZZ} \sin \beta_t \cos \beta_t \sin 2(\phi_t - \Phi)$$

$$q_{XZ} = \frac{3}{2}\hat{q}_{ZZ} \sin \beta_t \cos \beta_b \cos(\phi_t - \Phi)$$

$$q_{XX} - q_{YY} = \frac{3}{2}\hat{q}_{ZZ} \sin^2 \beta_t \cos 2(\phi_t - \Phi)$$

$$q_{ZZ} = \frac{1}{2}\hat{q}_{ZZ}(3 \cos^2 \beta_t - 1)$$

(5.23)

The quantities $\hat{p}_i$, $\hat{p}_{jk}$, $\hat{q}_i$, $\hat{q}_{jk}$ are the (coordinate-system independent) vector and tensor polarizations of beam and target as given by the occupation numbers of the hyperfine Zeeman states in a rotationally-symmetric frame along the z axis.

In the spin-correlation cross-section terms beam and target polarizations p and q appear as products. Therefore typical azimuthal dependences arise from combinations such as (and similarly for sin terms)

$$\propto \cos(\phi_b - \Phi) \cdot \cos(\phi_t - \Phi) \tag{5.24}$$

$$\propto \cos 2(\phi_b - \Phi) \cdot \cos(\phi_t - \Phi) \tag{5.25}$$

$$\propto \cos 2(\phi_b - \Phi) \cdot \cos 2(\phi_t - \Phi) \tag{5.26}$$

By using trigonometric relations it can be seen that these terms lead to azimuthal dependences $\propto [\cos M\Phi]_{M=0,1,...}$ and therefore to a maximum "complexity" of

$$\cos 4\Phi \text{ and also } \sin 4\Phi \tag{5.27}$$

e.g. for the correlation coefficient $C_{xy,xy}$. For this coefficient the product $\sin 2(\phi_b - \Phi) \cdot \sin 2(\phi_t - \Phi)$ may be transformed into

$$\frac{1}{2}(\cos 4\Phi + 1) \sin 2\phi_b \sin 2\phi_t$$

$$-\frac{1}{2}(\cos 4\Phi - 1) \cos 2\phi_b \cos 2\phi_t$$

$$-\frac{1}{2}\sin 4\Phi(\cos 2\phi_b \sin 2\phi_t + \cos 2\phi_t \sin 2\phi_b) \tag{5.28}$$

This complexity has to be met by a sufficiently fine-grained detector arrangement. It is clear that substantial simplifications arise with the choice of special polarization directions. If e.g. in an experiment both polarization vectors point in the x direction then with $\phi_b = \phi_t = 0$ only a simple ϕ dependence

$$\propto (1 - \cos 4\Phi) \tag{5.29}$$

of the cross section results for this correlation coefficient. Of course the different ϕ dependences have to be established for all coefficients. In Chap. 14 the complete set of terms entering the general cross-section of spin-1 on spin-1 correlations is given. For identical particles some terms are redundant.

5.5 Structure of the M Matrix and Number of "Necessary" Experiments

A nuclear reaction may be considered "completely measured", when all elements of the M matrix have been determined uniquely. How many independent polarization experiments will be necessary to achieve this goal depends on the spin structure of the reaction and can be derived theoretically. A number of investigations into this question has been published, see e.g. Hofmann and Fick [15], Fick [16], Köhler and Fick [17], Simonius [18] and the conclusions were that in the cases cited no measurement of e.g. four-spin observables was necessary. An example for an experimental program by which the "complete" determination of all M-matrix elements was attempted was the NN program at the Saclay SATURNE facility. At present the complete set of NN data is maintained by CNS DAC [19] e.g. for use with the phase-shift analysis program SAID.

Another approach consists in a least-squares fitting procedure of an incomplete set of data of certain reactions such as the two DD reactions at very low energies by a multi-channel R-matrix or single-channel T-matrix approach [14, 20–24].)

In practice—e.g. due to experimental uncertainties—some redundancy is used, i.e. more observables than minimally required will have to be measured. In addition not all M-matrix elements are independent since symmetries create relations between them:

- Rotational symmetry (conservation of angular momentum):
  The outcome of a measurement of a nuclear reaction is independent of the orientation of the coordinate system or of the orientation of an experiment in a given coordinate system.
- Mirror symmetry (Parity conservation—not for the weak interaction):
  The outcome of a measurement of a nuclear reaction is the same as that of a reaction reflected at the origin.
- Time-reversal symmetry: Since the time-reversal operator is "anti-unitary" there is no conserved quantity here, but relations exist between the observables of the forward and the backward reactions (as reversal of motion with suitable application to the spins of the particles involved).

Detailed investigations of the restrictions following for the M matrix can be found in Simonius [18], for special cases in Darden [5], Fick et al. [25]. Here only a few examples of practical importance will be given.

In principle the behavior of polarization observables under symmetry transformations can be derived by considering each of its components with respect to a given (Cartesian) coordinate system as a product of the corresponding tensor and the unit vector in the coordinate direction. Therefore not only the behavior of the spin tensor itself but also that of the coordinates have to be considered. In the helicity coordinate system the coordinates behave differently under the following two transformations by the parity operator P and time-reversal operator T

$$
\begin{array}{ccc}
 & x & y\ z \\
\text{Parity } \mathbf{P}: & -x & y\ -z,
\end{array}
$$

$$
\text{Time reversal } \mathbf{T}: x \quad y\ z \tag{5.30}
$$

i.e. only $\hat{y} = \dfrac{\vec{k}_{in} \times \vec{k}_{fin}}{|\vec{k}_{in} \times \vec{k}_{fin}|}$ is invariant under the parity transformation whereas all components of the Cartesian spin operator $\mathbf{S}$ are invariant. (This follows from the fact that $\mathbf{S}^2$ commutes with $\mathbf{P}$ and from the commutation relations $\mathbf{S} \times \mathbf{S} = i\mathbf{S}$). Therefore only S_y will be $\mathbf{P}$ invariant. This means that in a nuclear reaction with a parity-conserving interaction only the polarization component $p_{y'} \neq 0$ can be produced in the exit channel or the only component of an analyzing power $\neq 0$ will be A_y. On the other hand the measurement of components such as A_x or $A_z \neq 0$ is very suitable in the search for a parity violation.

From the parity behavior of $\mathbf{S}$ not only the parity behavior of higher-rank observables (for spins $> 1/2$) but also of the corresponding tensor moments can be obtained. Parity conservation imposes the condition on analyzing tensor moments (analyzing powers):

$$
T_{k-q} = (-1)^{k+q} T_{kq}. \tag{5.31}
$$

Like for spin-1/2 systems due to parity conservation the number of observables for larger spins will also be reduced. For $S = 1$ only one vector analyzing power A_y and the three tensor analyzing powers A_{zz}, A_{xz} and $A_{xx} - A_{yy}$ or iT_{11}, T_{20}, T_{21} and T_{22}, respectively can be $\neq 0$. Third-rank Cartesian analyzing powers thus

$$
A_{xxx} = A_{zzz} = A_{zzx} = A_{xxz} = A_{yyz} = A_{xyy} = 0. \tag{5.32}
$$

As a rule: the sum of the number of indices x and z must be even for the observable to exist [8].

For the generalized analyzing powers rotation invariance together with parity conservation yield the following relation [26]:

$$
\tau_{kqKQ}^{k'q'K'Q'}(p_3, \Theta_3, \phi_3; p_4, \Theta_4, \phi_4) = (-)^{\sum_{j=1}^{4} k_j} \left[\tau_{kqKQ}^{k'q'K'Q'}(p_3, \Theta_3, -\phi_3; p_4, \Theta_4, -\phi_4) \right]^*, \tag{5.33}
$$

which simplifies for the coordinate system introduced above in which $\phi_3 = 0$ and $\phi_4 = \pi$:

$$
\tau_{kqKQ}^{k'q'K'Q'} = (-)^{\sum_{j=1}^{4} k_j} \left[\tau_{kqKQ}^{k'q'K'Q'} \right]^* \tag{5.34}
$$

This means that all polarization-transfer coefficients are either real or imaginary. From the polarization of the outgoing particle 3

$$
t_{k'q'} = \frac{\tau_{0000}^{k'q'00}}{\tau_{0000}^{0000}}, \tag{5.35}
$$

also follows

$$t_{k'q'} = (-)^{k'} t^*_{k'q'}. \tag{5.36}$$

For $k' = 1$ (vector polarization) only Im $(t_{11}) = -\frac{1}{2}\sqrt{3}p_{y'}$ is $\neq 0$, i.e. the polarization vector points perpendicular to the scattering plane.

In Knutson [26], Ohlsen et al. [27] also the case of a reaction with three particles in the exit channel is discussed. In this case in general no restrictions of the number of observables by parity conservation apply. Exceptions are:

- The three particle momenta and the beam form a plane. The system then behaves like a two-particle reaction.
- Only one particle is detected (this is a kinematically incomplete measurement in which averaging over the momenta of the unobserved particles takes place). Here again the transfer coefficients will be either purely real or purely imaginary thus reducing their number by a factor 2.

Time-reversal invariance leads to relations between observables of the forward and the backward reaction. One such relation is the principle of "detailed balance" which requires equality (up to phase-space factors) of the cross-sections of both. For polarization observables similar relations result, e.g. the equality of the vector analyzing power for the forward reaction with polarized projectile a and the exit channel polarization of the ejectile b in the backward reaction produced with an unpolarized beam (or target)—which normally has to be measured in a second scattering. Since in elastic scattering projectile a and ejectile b are identical a double scattering experiment will yield A_y^2 absolutely (but not the sign of A_y). A widely used example for an analyzer reaction for protons is $^4He(p, p)^4He$.

Ohlsen [8] gives (for Cartesian observables) a detailed description of the formalism especially for the polarization-transfer and spin-correlation coefficients and (counting) rules for the restrictions imposed by parity conservation and time-reversal invariance. Systems with the spin structures

- Polarization transfer:

$$-\tfrac{\vec{1}}{2} + A \rightarrow \tfrac{\vec{1}}{2} + B$$
$$-\vec{1} + A \rightarrow \tfrac{1}{2} + B$$
$$-\tfrac{\vec{1}}{2} + A \rightarrow \vec{1} + B$$
$$-\vec{1} + A \rightarrow \vec{1} + B$$

- Spin correlations:

$$-\tfrac{\vec{1}}{2} + \tfrac{\vec{1}}{2} \rightarrow b \text{ or } B$$
$$-\vec{1} + \tfrac{\vec{1}}{2} \rightarrow b \text{ or } B$$
$$-\vec{1} + \vec{1} \rightarrow b \text{ or } B$$

Table 5.1 Table of the observables of a spin system $1/2 + 0 \longrightarrow 1/2 + 0$

Ejectile	Beam			
	1	p_x	p_y	p_z
$I(\Theta)$	$\begin{array}{c}\lvert A\rvert^2 + \lvert B\rvert^2\\ I_o\end{array}$		$\begin{array}{c}2\mathrm{Re}(A^*B)\\ I_oA_y\end{array}$	
$p_{x'}(\Theta)$		$\begin{array}{c}\lvert A\rvert^2 - \lvert B\rvert^2\\ I_oK_x^{x'}\end{array}$		$\begin{array}{c}2\mathrm{Im}\,(A^*B)\\ I_oK_z^{x'}\end{array}$
$p_{y'}I(\Theta)$	$\begin{array}{c}2\mathrm{Re}(A^*B)\\ I_op_y\end{array}$		$\begin{array}{c}\lvert A\rvert^2 + \lvert B\rvert^2\\ IK_y^{y'}\end{array}$	
$p_{z'}I(\Theta)$		$\begin{array}{c}-2\mathrm{Im}\,(A^*B)\\ I_oK_z^{x'}\end{array}$		$\begin{array}{c}\lvert A\rvert^2 - \lvert B\rvert^2\\ I_oK_z^{z'}\end{array}$

are discussed there. Sperisen et al. [10] contains a description of the general formalism for polarization-transfer experiments.

5.6 Examples

In the following only a few examples for the observables of important spin systems will be discussed.

5.6.1 Systems with Spin Structure $1/2 + 0 \longrightarrow 1/2 + 0$

Examples: $^4He(p, p)^4He\,^4He(n, n)^4He\,^1H(p, p)^1H$. The form of the transfer matrix for spin structure $1/2 + 0 \longrightarrow 1/2 + 0$ is

$$\mathbf{M}(\mathbf{k}_{in}, \mathbf{k}_{fin}) = A + B(\sigma\hat{\mathbf{y}}) \tag{5.37}$$

with A = non-spinflip amplitude, B = spinflip amplitude. The description will be in right-handed coordinate systems either in the c.m. or the lab. system. Table 5.1 shows the possible (and partly redundant) observables of these systems. For elastic scattering there is a substantial reduction of the number of independent observables by:

- Parity conservation (**P**): The expectation values of some observables vanish, e.g. the longitudinal analyzing power A_z;
- Time-reversal invariance (**T**): **T** connects observables of the forward and backward reaction.

With these symmetries the number of independent observables (experiments) is reduced from 15 possible to 3 independent "true" polarization experiments + the measurement of the differential and the total cross-sections (see Table 5.2) The

Table 5.2 Reduction of the number of independent observables by parity conservation (**P**) and time-reversal invariance (**T**)

	Beam unpolarized	Beam polarized
Differential cross-section	$\left(\frac{d\sigma}{d\Omega}\right)_0 (1)$	$A_i(3) \overbrace{(1)}^{\mathbf{P}}$
Ejectile polarization	$p_i(3) \overbrace{(1)}^{\mathbf{P}}$	$K_i^{j'}(9) \overbrace{(5)}^{\mathbf{P}} \overbrace{(2)}^{\mathbf{T}}$

number of complex amplitudes is $N = 2$, thus the number of real amplitudes is $2N - 1 = 3$, equal to the minimum number of necessary independent polarization experiments. Of these there are at most 4 (including the unpolarized cross-section) allowing for an additional relation between the observables. After identifying the transfer coefficients in the lab. system with the "Wolfenstein" parameters a well-known relation between different observables reads:

$$p_{y'}^2 + R^2 + A^2 = 1 \qquad (5.38)$$

5.6.2 Systems with the Spin Structure $1/2 + 1/2 \longrightarrow 1/2 + 1/2$

Examples are: the NN system, reactions such as $^3He(p, p)^3He$, $^3H(n, n)^3H$, etc. A very detailed description of the formalism of elastic scattering of these systems has been given in Bystricki et al. [2] where the connection between observables and M-matrix elements is made. Table 5.3 lists all possible polarization observables for this spin system. Thus in principle there are 255 possible polarization experiments + measurement of the unpolarized differential cross-section (+ measurement of the total cross-section). For elastic scattering parity conservation and time-reversal invariance will reduce this number to 25 for identical particles (such as in p–p scattering), and to 36 linear independent experiments for non-identical particles.

The form of the transfer matrix for systems with spin structure $1/2 + 1/2 \longrightarrow 1/2 + 1/2$: (elastic scattering) is

$$\mathbf{M}(\vec{k}_{in}, \vec{k}_{fin}) = 1/2[(a + b) + (a - b)(\sigma_1\mathbf{n})(\sigma_2\mathbf{n}) + (c + d)(\sigma_1\mathbf{m})(\sigma_2\mathbf{m})$$
$$+ (c - d)(\sigma_1\mathbf{l})(\sigma_2\mathbf{l}) + e(\sigma_1 + \sigma_2)\mathbf{n} \qquad (5.39)$$
$$+ f(\sigma_1 - \sigma_2)\mathbf{n}]$$

Here σ_i are the (Cartesian) Pauli spin operators and $\mathbf{m}, \mathbf{n}, \mathbf{l}$ the basis vectors of a right-handed c.m. coordinate system with:

$$\mathbf{l} = \frac{\vec{k}_{fin} + \vec{k}_{in}}{|\vec{k}_{fin} + \vec{k}_{in}|} \qquad \mathbf{m} = \frac{\vec{k}_{fin} - \vec{k}_{in}}{|\vec{k}_{fin} - \vec{k}_{in}|} \qquad \mathbf{n} = \frac{\vec{k}_{fin} \times \vec{k}_{in}}{|\vec{k}_{fin} \times \vec{k}_{in}|} \qquad (5.40)$$

Table 5.3 Observables for the spin structure $1/2 + 1/2 \longrightarrow 1/2 + 1/2$ with the designation of the classes of experiments as B for "beam", T for "target", u for "unpolarized", p for "polarized" are X_{pqik} with p,q,i,k, each with values s,n,l (polarized) or o (unpolarized) with the indices k for the target, i for the beam, q for the recoil nucleus and p for the ejectile. ℓ stands for "longitudinal", n for "normal" (along the scattering normal) and s (or m) for "sideways"(perpendicular to ℓ and n)

Observable	Bu,Tu	Bp,Tu	Bu,Tp	Bp,Tp
Differential cross-section	I_{oooo} (1)	A_{ooio} (3)	A_{oook} (3)	A_{ooik} (9)
Ejectile polarization	P_{pooo} (3)	D_{poio} (9)	K_{pook} (9)	M_{poik} (27)
Recoil polarization	P_{oqoo} (3)	K_{oqio} (9)	D_{oqok} (9)	N_{oqik} (27)
Polarization correlation	C_{pqoo} (9)	C_{pqio} (27)	C_{pqok} (27)	C_{pqik} (81)

In pp scattering, after considering parity conservation, time-reversal invariance and the Pauli principle there are $N = 5$, in np scattering $N = 6$ invariant, independent complex amplitudes (of a total of 16 possible ones).

Thus in a complete experiment $2N - 1$ real quantities have to be determined by at least as many independent experiments:

- for pp: 9
- for np: 11

5.6.3 The Systems with Spin Structure $\frac{1}{2} + \vec{1}$ and Three-Nucleon Studies

This system is in principle very important because the three-nucleon system $N + d$ is—after the NN system—the most important system for the test of fundamental interactions such as meson-exchange or effective-field theory NN input into Faddeev-like calculations, including tree-body and Coulomb forces. However, a very limited number of different polarization observables has been measured to date. These are nucleon and deuteron vector analyzing powers A_y and iT_{11}, as well as deuteron analyzing powers T_{2q} of elastic scattering, and analyzing powers of the breakup reactions $N + d \rightarrow 3N$. Observables of elastic scattering of the system: polarized spin-1/2 on unpolarized spin-1 particles including a phase-shift parametrization have been discussed in Ref. [28].

5.6.4 The Systems with Spin Structures $\vec{1} + \vec{1}$ and $\frac{1}{2} + \frac{1}{2}$ and the Four-Nucleon Systems

The four-nucleon system is the smallest nuclear system with a rich structure consisting of excited states and different channels with several clusterings $1 + 3$ as well as $2 + 2$. Its importance reaches from theoretical approaches using again the fundamental NN interactions via Faddeev-Yakubovsky equations and including three-nucleon, four-nucleon forces as well as the Coulomb interaction, to applications in fusion-energy research. Polarization effects in these reactions are relatively large, and a significant number of different observables has been collected, especially at low (fusion and astrophysically relevant) energies. In Sect. 14.2 the role of spin correlations of the D+D reactions in fusion energy will be discussed in detail.

5.6.5 Practical Criteria for the Choice of Observables

In practice sets of more experiments than minimally necessary are chosen for the following reasons:

- Consistency checks provided by relations between observables
- Resolution of possible discrete ambiguities caused by the bilinear form of the equations relating the M matrix with the observables
- Unavoidable experimental errors require that a fit procedure with more observables than fit parameters (matrix elements, phase shifts, etc.) is necessary.

Criteria for the selection of suitable polarization observables:

- Redundancy: observables should be linearly independent of each other (of course they depend on each other via different combinations of matrix elements)
- Technical realizability
- Availability e.g. of a polarized target
- Ease of orientation of the polarization in the beam and target into three orthogonal directions
- Avoidance of the complicated three or four-spin observables
- Avoidance of measuring a longitudinal polarization component in spin transfer (which needs spin rotation by magnetic field)
- "Sensitivity" of all observables to the amplitudes, small covariances between different observables (this may be important when determining reaction amplitudes in a fit procedure).

The measured (polarization) observables have to be compared to predictions of some (preferably the best available) theory for the description of a nuclear reaction. As an "interface" between theory and experiment a single observable could be used. When more (and different) observables have been measured it is better to extract common basic quantities such as the transfer-matrix elements, the S or T matrix elements or—for elastic scattering—a related parametrization such as phase shifts for comparison with the theory. This would also permit the prediction of unmeasured quantities from experimental data alone or in comparison with the theory. In the following the partial-wave analysis will be discussed.

References

1. Welton, T.A.: In: Marion J.B., Fowler, J.D. (eds.) Fast Neutron Physics II, p. 1317. Interscience, New York (1963)
2. Bystricki, J., Lehar, F., Winternitz, P.: J. de Physique **39**, 1 (1978)
3. Huber, P., Meyer, K.P. (eds.): Proceedings of the International Symposium on Polarization Phenomena of Nucleons, Basel 1960. Helv. Phys. Acta Suppl. VI, Birkhäuser, Basel (1961)
4. Barschall, H.H., Haeberli, W. (eds.): Proceedings of the 3rd International Symposium on Polarization Phenomena in Nuclear Reactions, Madison 1970, p. xxv. University of Wisconsin Press, Madison (1971)
5. Darden, S.E.: In: [6] p. 39 (1971)
6. Barschall, H.H., Haeberli, W. (eds.): Proceedings of the 3rd International Symposium on Polarization Phenomena in Nuclear Reactions, Madison 1970. University of Wisconsin Press, Madison (1971)
7. Petitjean, Cl., Huber, P., Paetz gen. Schieck, H., Striebel, H.R.: Helv. Phys. Acta **40**, 401 (1967)
8. Ohlsen, G.G.: Rep. Progr. Phys. **35**, 717 (1972)
9. Ohlsen G.G., Keaton, P.W. Jr: Nucl. Instrum. Methods **109**, 41 and ibid. p. 61 ff. (1973)
10. Sperisen, F., Grüebler, W., König, V.: Nucl. Instrum. Methods **204**, 491 (1983)
11. Sydow, L., Vohl, S., Lemaître, S., Nießen, P., Nyga, K.R., Reckenfelderbäumer, R., Rauprich, G., Paetz gen. Schieck, H.: Nucl. Instrum. Methods Phys. Res. A **327**, 441 (1993)
12. Sydow, L., Vohl, S., Lemaître, S., Patberg, H., Reckenfelderbäumer, R., Paetz gen. Schieck, H., Glöckle, W., Hüber, D., Witała, H.: Few-Body Systems **25**, 133 (1998)
13. Przewoski B., v. et al.: Phys. Rev. C **74**, 064003 (2006)
14. Paetz gen. Schieck, H.: Eur. Phys. J. A **44**, 321 (2010)
15. Hofmann, H.M., Fick, D.: Z. Physik **194**, 163 (1966)
16. Fick, D.: Z. Physik **199**, 309 (1967)
17. Köhler, W.E., Fick, D.: Z. Physik **215**, 408 (1968)
18. Simonius, M.: In: [29] p. 37 (1974)
19. CNS DAC, http://gwdac.phys.gwu.edu
20. Hale, G., Doolen, G.: Report LA-9971-MS, Los Alamos (1984)
21. Fletcher, K.A., Ayer, Z., Black, T.C., Das, R.K., Karwowski, H.J., Ludwig, E.J., Hale, G.M.: Phys. Rev. C **49**, 2305 (1994)
22. Lemaître, S., Paetz gen. Schieck, H.: Few-Body Systems **9**, 155 (1990)
23. Lemaître, S., Paetz gen. Schieck, H.: Ann. Phys. (Leipzig) **2**, 503 (1993)
24. Geiger, O., Lemaître, S., Paetz gen. Schieck, H.: Nucl. Phys. **A586**, 140 (1995)
25. Fick, D., Einf. i. d. Kernphysik m. polarisierten Teilchen, BI-HTB 755/755a, Mannheim (1971)
26. Knutson, L.D.: Nucl. Phys. **A198**, 439 (1972)
27. Ohlsen, G.G., Brown, R.E., Correll, F.D., Hardekopf, R.A.: Nucl. Instrum. Methods **179**, 283 (1981)
28. Seyler, R.G.: Nucl. Phys. **A124**, 253 (1969)
29. Fick, D. (ed.): Proceedings of Meeting on Polarization Nuclear Physics, Ebermannstadt 1973, Lecture Notes in Physics, vol. 30. Springer, Berlin (1974)

Chapter 6
Partial Wave Expansion

Especially at low energies the partial-wave expansion of the observables is useful. One advantage is that—since the Legendre functions are eigenfunctions of the angular momentum—the influence of and dependence on different angular momenta in the reaction can be studied. When dealing with the nuclear part of the interaction— due to the short range of nuclear forces—the expansion can be truncated after a few terms; the centrifugal barrier prevents higher angular momenta from contributing. The problem with incident charged particles is that the Coulomb interaction, due to its long range, requires a very large number of partial waves.

6.1 Neutral Particles

The most general expansion for two-particle reactions between neutral particles was published by Welton [1]. It describes the (spherical) tensor moments of the exit channel as function of the tensor moments of the entrance channel, as prepared. By Heiss [2] it was extended to elastic scattering of charged reaction partners and by Hofmann, Aulenkamp, Nyga in addition to the case of identical particles. Here the final result of Welton will be given with the modification that the definition of the tensor moments follows that of Lakin [3] and therefore complies with the *Madison convention* [4]. In order to avoid confusion with expressions of the R-matrix theory [5], here the R and $\mathcal{R}$ of Welton have been renamed T and $\mathcal{T}$, the tensor moments are designated as introduced in the present text: $t^{kq,KQ}$ instead of $t_{q\gamma,Q\Gamma}$ for the exit channel, $t_{k'q',K'Q'}$ instead of $t_{q'\gamma',Q'\Gamma'}$ for the entrance channel.

$$t^{kq,KQ} = (2k_{in})^{-2}(\hat{i}\hat{I})^{1/2}$$

$$\cdot \sum \begin{Bmatrix} i & I & s_1 \\ k & K & t \\ i & I & s_2 \end{Bmatrix} \begin{Bmatrix} i' & I' & s'_1 \\ k' & K' & t' \\ i' & I' & s'_2 \end{Bmatrix} \begin{Bmatrix} l_1 & s_1 & J_1 \\ l & t & L \\ l_2 & s_2 & J_2 \end{Bmatrix} \begin{Bmatrix} l'_1 & s'_1 & J_1 \\ l' & t' & L \\ l'_2 & s'_2 & J_2 \end{Bmatrix}$$

$$\cdot (l_1 l_2 00 | l0)(l'_1 l'_2 00 | l'0)(l t 0\Lambda | L\Lambda)$$

H. Paetz gen. Schieck, *Nuclear Physics with Polarized Particles,*
Lecture Notes in Physics 842, DOI: 10.1007/978-3-642-24226-7_6,
© Springer-Verlag Berlin Heidelberg 2012

$$\cdot (l't'0\Lambda'|L\Lambda')(kKqQ|t\Lambda)(k'K'q'Q'|t'\Lambda')$$

$$\cdot T^{J_1^{\pi_1}} T^{J_2^{\pi_2}*} D_{\Lambda'\Lambda}^L(\Phi,\Theta,0)$$

$$\cdot (\hat{\imath}'\hat{I}')^{-1/2}$$

$$\cdot t_{k'q',K'Q'} \tag{6.1}$$

Meaning of the notation:

- Primed quantities : entrance channel, unprimed ones: exit channel
- Alternatives in each channel are distinguished by 1 and 2
- Particle spins: i, I, i', I'
- Channel spins: s, s'
- Orbital angular momenta: ℓ, ℓ', total angular momentum: J (the only conserved angular momentum)
- Rank and component of the tensor moments: k, q and K, Q
- Sums are over all indices except $k, K, q, Q; I, i, I', \imath'$
- $\hat{I}$ means $2I+1$
- Symbols in wavy brackets are the 9j symbols (see [6])
- Symbols like $(l_1 l_2 00|l0)$ are the Clebsch–Gordan coefficients (see [6])
- The matrix elements $T^{J_1^{\pi_1}}$ and $T^{J_2^{\pi_2}}$, respectively, are defined in a representation with the asymptotically good quantum numbers as:

$$T^{J_1^{\pi_1}} = \langle \alpha'\ell's'|T|\alpha\ell s\rangle, \tag{6.2}$$

where $T = S - 1$ defined in spin space with S being the usual S matrix. Already from the abbreviated form:

$$t \propto \sum_{1,2} B(1,2)T_1 T_2^* D_{\Lambda\Lambda'}^L \cdot t', \tag{6.3}$$

some general conclusions can be derived.
$B(1,2)T_1 T_2^* D_{\Lambda\Lambda'}^L$ are components of the generalized analyzing powers $T_{k'q',K'Q'}^{kq,KQ}$. By interchanging indices $1 \leftrightarrow 2$ one finds that $B(2,1) = (-)^{k+K+k'+K'} \cdot B(2,1)$ and

$$t \propto \sum_{1,2} \frac{1}{2}[T_1 T_2^* B(1,2) + T_1^* T_2 B(2,1)]D_{\Lambda'\Lambda}^L \cdot t'$$

$$= \sum_{1,2} 1/2(T_1 T_2^* B(1,2) + (-)^{k+K+k'+K'} T_1^* T_2 B(1,2)D_{\Lambda\Lambda'}^L \cdot t' \tag{6.4}$$

For

$$k + K + k' + K' = \begin{Bmatrix} even \\ odd \end{Bmatrix} \quad only \quad \begin{Bmatrix} \mathrm{Re} \\ \mathrm{Im} \end{Bmatrix} (T_1 T_2^*) \tag{6.5}$$

will appear. If e.g. only the incident *beam* is polarized and no outgoing polarization is measured ($k' = 1, k = K = K' = 0$), the analyzing power is

$$A_y \propto i T_{11} \propto \mathrm{Im}\,(T_1 T_2^*).\tag{6.6}$$

Thus polarization effects of odd rank (e.g. the vector analyzing power or the vector polarization) vanish if

- the matrix elements are purely real. This will be the case e.g. in Born approximation with a real potential;
- only a single matrix element contributes: $T_1 T_1^* - T_1^* T_1 = 0$. This is the case for an isolated resonance with only one value of the orbital angular momentum (if no tensor force couples angular momenta of equal parity) and with no direct background contribution. An example is the $3/2^+$ resonance of the ^{3}H(d,n)^{4}He reaction at $E_d = 107\,\mathrm{keV}$;
- all matrix elements have the same phase: with $T_1 = r_1 e^{i\phi}, T_2 = r_2 e^{i\phi}$: $T_1 T_2^* = r_1 r_2 = \mathrm{real}$;
- only one value of ℓ exists and is zero;
- only one intermediate state with $J_1 = J_2 = 0$ or $1/2$ exists. In the last two cases the angular distribution of the unpolarized cross-section is isotropic: from $\ell_1' = \ell_2' = \ell = \ell + t(= 0) = L = 0 \to \sigma_0$ is isotropic;
- there is no interaction distinguishing (for one ℓ) between the two possible different values of J. A vector $\vec{\ell} \cdot \vec{s}$ force is e.g. necessary for producing vector polarization or analyzing power, resp., otherwise the above condition $T_1 \neq T_2$ is not fulfilled.

Additional conclusions:

- Parity conservation reduces the number of possible tensor moments.

Example The outgoing tensor moment $t_{00} = 0$ with incident tensor moment t_{10}', polarized in the z direction, and therefore the (longitudinal) analyzing power of a parity conserving reaction $A_z \propto T_{10} = 0$. This is due to the one property of the CG coefficient $(\ell_1' \ell_2' 00 | \ell' 0)$ which is only $\neq 0$ if $\ell_1' + \ell_2' + \ell'$ is even, and analogously for $(\ell_1 \ell_2 00 | \ell 0)$. Parity conservation requires that $\ell_1' + \ell_2' + \ell_1 + \ell_2$ be even, resulting in $L = \ell$ and $\ell + \ell' = $ even. However, with $k' = 1, q' = 0, k = q = 0$ and therefore $t' = 1, L = \ell' + t'$ we have $\ell = \ell_1' \pm 1$ and $\ell + \ell' = 2\ell' \pm 1 = $ odd in contradiction to the above.

- The complexity (defined as the maximum possible order L of the functions $D_{\Lambda'\Lambda}^L$ or Y_L^Λ or P_L^Λ) of angular distributions can be obtained as:

  - $L_{\max} \leq J_1 + J_2$
  - $L_{rmmax} \leq \ell_1 + \ell_2 + k + K$
  - $L_{\max} \leq \ell_1' + \ell_2' + k' + K'$

- It is evident that for the unpolarized cross-section $t_{00,00} \Lambda = \Lambda' = 0$, i.e. there is no Φ dependence and D_{00} is reduced to a simple Legendre polynomial $P_L(\cos\Theta)$. Especially for s waves ($\ell_1 = \ell_2 = L = 0$) the angular distribution becomes isotropic. This also holds for $J_1 = J_2 = 0$ or $= 1/2$. Inspection of the relevant J symbol shows that, as in the case of $J_1 = J_2 = 1/2$, $\ell = L = 0$ is the maximum possible value:

$$
\begin{Bmatrix} \ell_1 & s_1 & J_1 \\ \ell & t & L \\ \ell_2 & s_2 & J_2 \end{Bmatrix} = \begin{Bmatrix} \ell_1 & s_1 & 0 \\ \ell & 0 & \ell \\ \ell_2 & s_2 & 0 \end{Bmatrix} \tag{6.7}
$$

6.2 Charged Particles

The case including the Coulomb interaction in elastic scattering has been treated by Heiss [2] in such a way that instead of one expression for an observable there are now three: the pure nuclear-interaction term, the pure Rutherford term, and an interference term between both. This last one is—due to the long range of the Coulomb force—the one which may cause problems when truncating higher partial waves where they should be included up to very high ℓ values (corresponding to a large screening (or cut-off) radius for the interactions) while for the pure nuclear term very few low-ℓ partial waves suffice. The pure Coulomb term is written down in closed form—it is just the Rutherford scattering, at least when dealing with the monpole term of the Coulomb force, i.e. between point charges.

The most general equation relating outgoing tensor moments with incident ones for the scattering of charged particles thus has three parts.

$$
\begin{aligned}
t^{kq,KQ} = (2k_{in})^{-2} \Big\{ & 4\pi \delta_{\alpha\alpha'} \delta_{iI,i'I'} |C_\alpha(\Theta)|^2 \sum B_1(kqKQ; k'q'K'Q'; L\Lambda\Lambda') \\
& + (4\pi)^{1/2} \delta_{\alpha\alpha'} \delta_{iI,i'I'} \sum B_2(\ell s_2 kqKQ; \ell' s_2' k'q'K'Q'; L\Lambda\Lambda'; I) \\
& \cdot \Big[iC(\Theta)T^* + (-)^{k+k'+K+K'} \big(iC(\Theta)T^* \big)^* \Big] \\
& + \frac{1}{2} \sum B_4(\ell_1 s_1 \ell_2 s_2 kqKQ; \ell_1' s_1' \ell_2' s_2' k'q'K'Q'; L\Lambda\Lambda'; J_1 J_2) \\
& \cdot \Big[(T_1 T_2^*) + (-)^{k+k'+K+K'} (T_1 T_2^*)^* \Big] \Big\} D_{\Lambda\Lambda'}^L(\Phi, \Theta, 0) t_{k'q',K'Q'}
\end{aligned}
\tag{6.8}
$$

The sums run over all arguments of the B coefficients; the B coefficients are defined as the Rutherford term

$$
B_1 = \delta_{kk'} \delta_{KK'} (kKq'Q'|L\Lambda)(kKqQ|L\Lambda'), \tag{6.9}
$$

the interference term

$$B_2 = (-)^{s_2+s'_2-2I} \hat{I}\hat{k}'\hat{K}'(\hat{\ell}\hat{\ell}'\hat{s}_2\hat{s}'_2)^{1/2}$$

$$\cdot \sum_{tt's} (\hat{t}\hat{t}')^{1/2}\hat{s} \begin{Bmatrix} i & I & s \\ k & K & t \\ i & I & s_2 \end{Bmatrix} \begin{Bmatrix} i & I & s \\ k' & K' & t' \\ i & I & s' \end{Bmatrix} W(st I\ell; s_2 L) W(st' I\ell'; s'_2 L)$$

$$\cdot (t\ell\Lambda'0|L\Lambda')(t'l'\Lambda0|L\Lambda)(kKqQ|t\Lambda')(k'K'q'Q'|t'\Lambda),$$

$$\tag{6.10}$$

and the pure nuclear term (the Welton formula)

$$B_4 = \left(\frac{\hat{i}\hat{I}}{\hat{k}\hat{K}}\right)^{1/2} \left(\frac{\hat{i}'\hat{I}'}{\hat{k}'\hat{K}'}\right)^{1/2}$$

$$\cdot F(\ell_1\ell_2 s_1 s_2 J_1 J_2 L\Lambda'; kKqQ) F(\ell'_1\ell'_2 s'_1 s'_2 J_1 J_2 L\Lambda; k'K'q'Q') \tag{6.11}$$

with

$$F(\ell_1\ell_2 s_1 s_2 J_1 J_2 L\Lambda; kKqQ) = (\hat{\ell}_1\hat{\ell}_2\hat{s}_1\hat{s}_2\hat{J}_1\hat{J}_2\hat{k}\hat{K})^{1/2}(-)^{\ell_1+L}$$

$$\cdot \sum_{\ell t} (l_1 l_2 00|\ell 0)(lt0\Lambda|L\Lambda)(kKqQ|t\Lambda)$$

$$\tag{6.12}$$

$$\cdot \begin{Bmatrix} i & I & s_1 \\ k & K & t \\ i & I & s_2 \end{Bmatrix} \begin{Bmatrix} \ell_1 & s_1 & J_1 \\ \ell & t & L \\ \ell_2 & s_2 & J_2 \end{Bmatrix}$$

The quantities in these equations are as in Eq. 6.1. In addition,

$$C(\Theta) = (4\pi)^{-1/2}\eta \csc^2\left(\frac{\Theta}{2}\right) \exp\left\{-2i\eta \ln\left[\sin\left(\frac{\Theta}{2}\right)\right]\right\} \tag{6.13}$$

is the Coulomb (Rutherford) amplitude describing the long-range part of the interaction, and the W coefficients are Racah coefficients equivalent to 6j symbols (see [6]).

6.3 Computer Codes

The formalism of Welton/Heiss has been transformed repeatedly into computer programs:

- TENMO at Oak Ridge (ORNL-4125)
- FATSO (Seiler, Basel) [7]
- FATSON (Seiler, Aulenkamp, Cologne) [7]
- TUFO (Aulenkamp): [8]
- TUFX (Aulenkamp)
- TUFXID and FATSONID (Hofmann, H.M., Erlangen), [9] for identical particles in the entrance channel

- Specialized to spin-structure spin-1 on spin-1: TUFXDD and DD (Lemaître, Geiger, Cologne); CORPOL (Ad'yasevich et al., Moscow)

These programs enabled the user to:

- obtain the forefactors of the sums of products of the complex matrix elements in the complete matrix-element expansion;
- predict any observables provided values of the matrix elements were given;
- obtain a least-squares fit of predictions to measured observables;
- check the quality of the fit with the input observables.

Such a project has been carried out for the two $D + D$ reactions at energies below 1.5 MeV (0.5 MeV for the $D(d,n)^3He$ reaction) and has led to predictions for unobserved quantities such as spin correlations and the quintet-suppression factors, important for the possible suppression of unwanted neutrons in fusion-energy applications (see Chap. 14 below).

References

1. Welton, T.A.: In: Marion J.B., Fowler, J.D. (eds.) Fast Neutron Physics II, p. 1317. Wiley Interscience, New York (1963)
2. Heiss, P.: Z. Physik **251**, 159 (1972)
3. Lakin, W.: Phys. Rev. **98**, 139 (1958)
4. Barschall, H.H., Haeberli, W. (eds.): In: Proceedings of the 3rd Internationl Symposium on Polarization Phenomena in Nuclear Reactions, Madison 1970, p. xxv. University of Wisconsin Press, Madison (1971)
5. Lane, A.M., Thomas, R.G.: Rev. Mod. Phys. **30**, 257 (1958)
6. Brink, D.M., Satchler, G.R.: Angular momentum. Oxford University Press, Oxford (1971)
7. Seiler, F.: Comp. Phys. Comm. **6**, 229 (1974)
8. Aulenkamp, H.: Diploma thesis, Universität zu Köln (1973, unpublished)
9. Nyga, K.: Diploma thesis, Universität zu Köln (1985, unpublished)

Chapter 7
Charged-Particle Versus Neutron-Induced Reactions

The questions of charge symmetry and charge independence (isospin conservation) and their possible breaking have always been important, also as a field where mass differences between up and down quarks in the nucleons might show up. Examples for such effects are the Nolen-Schiffer anomaly of the Coulomb energy differences of mirror nuclei [1] and the difference between the scattering lengths a_{nn}, a_{pp}, and a_{np} of the singlet $^1S_0$ nucleon-nucleon interaction (for a recent discussion see Ref. [2]). When initiating nuclear reactions with protons and comparing them to their neutron-induced mirror reaction, after correctly subtracting the "trivial" Coulomb part of the interaction, under isospin conservation the remaining observables should be equal. This is of course true for all, i.e. also polarization observables. Therefore, the use of polarized neutrons which normally have to be produced in special nuclear reactions (see Sect. 10.2) is very important. Examples are the intensive study of the three-particle breakup reaction ^{2}H(p,pp)n and comparison with ^{2}H(n,nn)^{1}H, as well as of the elastic scatterings ^{2}H(p,p)^{2}H and ^{2}H(n,n)^{2}H. Although neutron-induced reactions are technically more difficult and in general less precise than their proton-induced counterparts they are important because the realistic inclusion of the long-range Coulomb interaction into "numerically exact" Faddeev calculations has been achieved only recently allowing now the realistic study of charged-particle three-body reactions. The comparison between both revealed not only discrepancies between them, but of both with most advanced theories such as meson exchange or EFT Faddeev-type calculations (for recent discussions of these low-energy discrepancies see e.g. Refs. [3, 4]).

Another way to "see" isospin violations is to look for observables of isospin-forbidden reactions. A recent example is the cross-section found to be $\neq 0$ [5] of the reaction

$$d + d \rightarrow \alpha + \pi^0. \tag{7.1}$$

Polarization observables have been investigated in isospin-forbidden reactions such as the deuteron breakup reaction ^{4}He($\vec{d}$, pα)n in the kinematical configuration

H. Paetz gen. Schieck, *Nuclear Physics with Polarized Particles*,
Lecture Notes in Physics 842, DOI: 10.1007/978-3-642-24226-7_7,
© Springer-Verlag Berlin Heidelberg 2012

including maximum np final-state interaction, i.e. with the np relative energy $E_{np} = 0$. The spin–isospin situation is:

$$
\begin{array}{llcccc}
 & & & \overset{\text{d*}}{\overbrace{}} & \\
\text{Reaction} & \vec{\text{d}} + {}^4\text{He} & \rightarrow & \text{pn} & + \alpha \\
\text{Spin} & 1 & 0 & 0 & 0 \\
\text{Isospin} & 0 & 0 & 1 & 0.
\end{array}
$$

Here the transition from the (triplet) deuteron d (isospin 0) to the singlet deuteron d* (isospin 1 due to the Pauli principle) is isospin-forbidden. The spin structure of the reaction is such that the (transverse) tensor analyzing power A_{yy} in this kinematical configuration is -1, independent of energy and angle [6] and is therefore a sensitive indicator of d* production. Indications of isospin breaking have been found, see Refs. [7, 8].

References

1. Nolen, J.A. Jr., Schiffer, J.P. : Ann. Rev. Nucl. Sci. **19**, 471 (1969)
2. Šlaus, I.: Nucl. Phys. **A790**, 199c (2007)
3. Tornow, W., Esterline, J.H., Weisel, G.J.: Nucl. Phys. **A790**, 64c (2007)
4. Sagara, K.: Few-Body Systems. **48**, 59 (2010)
5. Stephenson, E.J., Bacher, A.D., Allgower, C.E., Gårdestig, A., Lavelle, C.M., Miller, G.A., Nann, H., Olmsted, J., Pancella, P.V., Pickar, M.A., Rapaport, J., Rinckel, T., Smith, A., Spinka, H.M., van Kolck, U.: Phys. Rev. Lett. **9**, 142302 (2003)
6. Jacobsohn, B.A., Ryndin, R.M.: Nucl. Phys. **24**, 505 (1961)
7. Gaiser, N.O., Darden, S.E., Luhn, R.C., Paetz gen. Schieck, H., Sen, S.: Phys. Rev. C **38**, 1119 (1988)
8. Niessen, P., Lemaître, S., Nyga, K.R., Rauprich, G., Reckenfelderbäumer, R., Sydow, S., Paetz gen. Schieck, H., Doleschall, P.: Phys. Rev. C **45**, 2570 (1992)

Part III
Devices

Chapter 8
Sources and Targets of Polarized H and D Ions

8.1 Physical Basics: General Introduction

The Stern–Gerlach experiment [1] showed for the first time the quantization of angular momentum, more precisely the quantization of a new degree of freedom later called electron spin.[1] The observation consisted in the spatial splitting of an Ag atomic beam in an inhomogeneous magnetic field (Fig. 8.1). The classical expectation was that the energy of a magnetic dipole (a magnetic moment) would behave according to a continuous energy distribution

$$W = -\vec{\mu}\vec{B} = -\mu B \cos\theta \tag{8.1}$$

and that such a magnet would undergo a force in the inhomogeneous magnetic field

$$\vec{F} = -\vec{\nabla}W = -\mu\vec{\nabla}|\vec{B}| \tag{8.2}$$

Surprisingly in the Stern–Gerlach experiment two discrete energy values appeared causing a deflection into an upper and a lower Ag spot: a quantum-mechanical dipole behaved completely different from a classical one.

This behavior is best described by looking at the energy of the magnetic moment in the magnetic field [7]. By measuring the energy of such a system in the magnetic field one finds it proportional to B. The proportionality coefficient is called the component of m_J in the direction of the field ("direction of quantization"). The unit of measurement is one Bohr magneton or one nuclear magneton μ_B or μ_N, respectively. Thus the magnetic moments belonging to the electronic spin J or nuclear spin I are

---

[1] This interpretation was, however, only put forward in 1927 [2], 2 years after Goudsmit and Uhlenbeck found evidence of the half-valued spin of the electron [3]. The delay can be attributed to the fact that Stern and Gerlach measured the size of the magnetic moment of the Ag atom to be about $1\mu_B$ [4] appropriate for the orbital angular momentum $1\hbar$—but with a third component missing. On the other hand a factor 2 compensating for the factor 1/2 of spin-$\frac{1}{2}\hbar$ (the "Thomas factor" [5]) was not known at that time, see also Friedrich and Herschbach [6].

H. Paetz gen. Schieck, *Nuclear Physics with Polarized Particles*,
Lecture Notes in Physics 842, DOI: 10.1007/978-3-642-24226-7_8,
© Springer-Verlag Berlin Heidelberg 2012

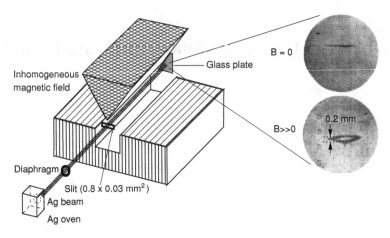

Fig. 8.1 Setup and result of the classic Stern–Gerlach experiment [1]

$$\mu_J = g_J \mu_B J \quad \text{or} \quad \mu_I = g_I \mu_N I \tag{8.3}$$

with (in MKS units)

$$\mu_B = \frac{e\hbar}{2m_0} = 9.27 \cdot 10^{-24} \text{J/T} \quad ; \quad \mu_N = \frac{e\hbar}{2m_p} = 5.05 \cdot 10^{-27} \text{J/T}, \tag{8.4}$$

where m_0 and m_p are the electron and proton masses, respectively. There is an important *convention*: For positively charged particles and with $g > 0$ spin and magnetic moment vectors are parallel, for negatively charged particles (e.g. electrons) they are antiparallel. The g are the (structure-dependent) g factors (often called Landé factors). Another quantity, the "gyromagnetic ratio" γ is defined generally as

$$\mu_s = \gamma_s S \tag{8.5}$$

The energy of a spin state S in a magnetic field B is:

$$W = -\mu B = -\mu \left(\frac{S_z}{S} \right) B \quad \text{(classical)} \quad W = g_S \mu_B S\hbar B \quad \text{(q.m., resp.)} \tag{8.6}$$

A system with spin S in a magnetic field splits into $2S+1$ components (=possibilities of orientation) with magnetic quantum numbers $-S \leq m_S \leq S$. In the case of hydrogen atoms this is the fine-structure ("FS") Zeeman effect. This is what is observed in the Stern–Gerlach experiment. The forces acting on each of these components in an inhomogeneous magnetic field are different and, especially, dependent on the sign of m_S. This is the principle of a "spin filter" for the spatial separation of spin states. For practical applications e.g. for sources of polarized atoms/ions where high beam intensities are desired the Stern–Gerlach arrangement with separation in one dimension has been replaced by systems with rotational symmetries, i.e. multipole fields. With these and diaphragms for the elimination of the wrong components and transmission of the wanted state we learn that a Stern–Gerlach filter acts as a nearly perfect polarizer. As an example, the component with $m_J = +\frac{1}{2}$ of the electron spin J of atoms in an infinitely strong magnetic field is ~100% polarized in J. This is true as long as the particle velocities in the beam are low enough (e.g. thermal) that the existing magnets can separate the two spin components completely, and the widths of the partial beams are small enough to be separable (the widths are a consequence of the atoms' velocity distributions and therefore a function of temperature).

It is interesting that with such systems of spin separation magnets combined with diaphragms and beam stops Feynman in his Lectures [7] (see also [8]) introduced the projector formalism setting up elementary properties of quantum systems.

Because of the smallness of the nuclear μ_I with the nuclear spin $I \neq 0$ the separation of the spin components is more complicated and one has to make use of the *hyperfine structure (HFS)* and its Zeeman effect. Here we have a system of two spins (nuclear spin i and electronic spin J) coupled together to $\vec{F} = \vec{I} + \vec{J}$ in a magnetic field. Two limiting cases are:

- Very weak magnetic field $B \to 0$. $\vec{I}$ and $\vec{J}$ are strongly coupled to $\vec{F}$. F and m_F are good quantum numbers and $m_F = m_I + m_J$, $|I - J| \leq F \leq I + J$. The energy eigenvalues behave like with the FS, i.e. proportional to the field B.
- Very strong magnetic field $B \to \infty$ ($\mu_B B \gg \Delta W_{HFS}$). $\vec{I}$ and $\vec{J}$ are decoupled, F is not a good quantum number, but instead I, J, μ_I, and μ_J are (*Paschen-Back effect*). Because of $\mu_J \gg \mu_I$ the energy splitting is basically given by $\Delta W_J = g_J \mu_B (m_J/\hbar) B$, and $\Delta W_I = g_I \mu_N (m_I/\hbar) B$ is only a small splitting correction.
- For the general case in intermediate magnetic fields the Schrödinger equation for the coupled system has to be solved explicitly.

8.2 Hyperfine Structure

8.2.1 HFS of the H Atom

Figure 8.2 shows the energy levels of the hydrogen atom. The HFS is generated by the coupling of the electronic spin $J = \frac{1}{2}$ with the nuclear spin $I = \frac{1}{2}$ (resp. $I = 1$ for deuterium) to a total angular momentum $\vec{F} = \vec{I} + \vec{J}$ with the state vectors $|F, m_F\rangle$. The eigenvalue equation reads:

$$H_{HFS}|F, m_F\rangle = E_F|F, m_F\rangle, \tag{8.7}$$

where[2]

$$H_{HFS} = a(\vec{I}\vec{J}) \tag{8.8}$$

---

[2] Smaller and different terms such as the quadrupole interaction in the HFS hamiltonian have been neglected here.

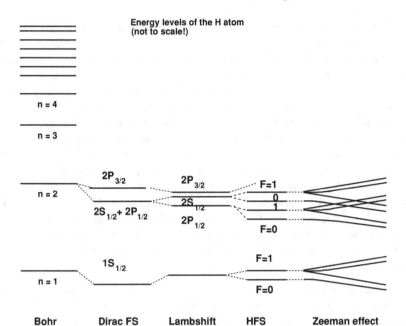

Fig. 8.2 Energy levels of the H atom (not to scale!)

With $\vec{I}\vec{J} = \frac{1}{2}[F(F+1) - I(I+1) - J(J+1)]$ follows:

$$E_F = \frac{a\hbar^2}{2}\left[F(F+1) - I(I+1) - \frac{3}{4}\right] \tag{8.9}$$

and the HFS splitting becomes:

$$\Delta W = E_{F=I+\frac{1}{2}} - E_{F=I-\frac{1}{2}} = a\hbar^2\left(I+\frac{1}{2}\right) = a\hbar^2 \begin{cases} \cdot\frac{3}{2} & \text{for D} \\ \cdot 1 & \text{for H} \end{cases} \tag{8.10}$$

8.2.2 HFS in a Magnetic Field (Zeeman Effect)

For an arbitrary intermediate magnetic field it is useful to expand into states with "good" quantum numbers, i.e. those either in a very weak or an very strong magnetic field. Here—because we are interested in the nuclear polarization—one preferably expands into $|Im_I\rangle|Jm_J\rangle \equiv |m_J, m_I\rangle$. Following the rules of vector coupling we obtain:

$$m_F = m_I + m_J \quad \text{and} \quad |m_J, m_I\rangle = |m_J, m_F - m_J\rangle \tag{8.11}$$

especially for $m_J \pm \frac{1}{2}$: $|m_J, m_I\rangle = |\frac{1}{2}, m_F \mp \frac{1}{2}\rangle$. The total Hamiltonian reads:

$$H = H_{nuclear} + H_{electronic} + (\gamma_I m_I + \gamma_J m_J)B_0 + a(\vec{I}\vec{J}) \quad (+H_{quad.}) \quad (8.12)$$

The following relations are useful here:

$$\vec{I}\vec{J} = I_z J_z + \frac{1}{2}(I_+ J_- + I_- J_+) \quad \text{and} \tag{8.13}$$

$$\langle jm \pm 1|J_\pm|jm\rangle = [(j \pm m + 1)(j \mp m)]^{1/2}. \tag{8.14}$$

Thus we obtain for

$$H = \begin{pmatrix} H_{11} & H_{12} \\ H_{21} & H_{22} \end{pmatrix} \quad \text{the matrix elements:} \tag{8.15}$$

$$H_{11} = \langle \frac{1}{2}, m_F - \frac{1}{2}|H|\frac{1}{2}, m_F - \frac{1}{2}\rangle = \frac{1}{2}\left[(\gamma_J + \gamma_I)B\hbar + a\hbar^2\left(m_F - \frac{1}{2}\right)\right]$$

$$= \frac{\Delta W}{2}\left[(\gamma_J + \gamma_I)\frac{B\hbar}{\Delta W} + \frac{(m_F - \frac{1}{2})}{(I + \frac{1}{2})}\right] \tag{8.16}$$

$$H_{22} = \langle -\frac{1}{2}, m_F + \frac{1}{2}|H| - \frac{1}{2}, m_F + \frac{1}{2}\rangle = -\frac{1}{2}\left[(\gamma_J + \gamma_I)B\hbar + a\hbar^2\left(m_F + \frac{1}{2}\right)\right]$$

$$= -\frac{\Delta W}{2}\left[(\gamma_J + \gamma_I)\frac{B\hbar}{\Delta W} + \frac{(m_F + \frac{1}{2})}{(I + \frac{1}{2})}\right] \tag{8.17}$$

$$H_{12} = H_{21} = \frac{a\hbar^2}{2}\sqrt{(I + \frac{1}{2})^2 - m_F^2} = \frac{\Delta W}{2}\sqrt{\left(I + \frac{1}{2}\right)^2 - m_F^2}/(I + \frac{1}{2}) \tag{8.18}$$

The eigenvalues are the roots of the secular equation:

$$\begin{vmatrix} H_{11} - \lambda_\pm & H_{12} \\ H_{12} & H_{22} - \lambda_\pm \end{vmatrix} = 0, \quad \text{i.e.} \quad \lambda_\pm^2 - \lambda_\pm(H_{11} + H_{22}) + (H_{11}H_{22} - H_{12}^2) = 0 \tag{8.19}$$

with the *Breit-Rabi equation* [9] as solution:

$$\lambda_\pm = E_{F=I\pm\frac{1}{2}} = -\frac{\Delta W}{2}\left[\frac{1}{2I + 1} - \underbrace{\frac{\gamma_I B m_F}{\Delta W/2}}_{small} \pm \sqrt{1 + \frac{4m_F}{2I + 1}x + x^2}\right] \tag{8.20}$$

with $x = (g_J - g_I)B/\Delta W \approx g_J B/\Delta W = B/B_{crit}$. For practical purposes often terms with g_I or γ_I, resp., may be neglected due to the smallness of the nuclear magnetic moment in comparison with the electronic moment.

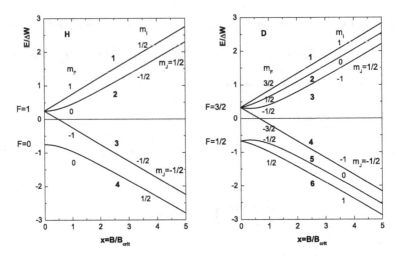

Fig. 8.3 Zeeman splitting (Breit-Rabi diagrams) of the HFS of a system with $J = 1/2$; $I = 1/2$ (e.g. H) and a system with $J = 1/2$; $I = 1$ (e.g. D) as function of the magnetic-field parameter $x = B/B_{crit}$. The states are usually numbered consecutively starting with the highest-energy state

The parameter x is the magnetic field (dimensionless), measured in units of the magnetic field corresponding to the HFS splitting $B_{crit} = \Delta W/(g_J - g_I)\hbar \approx \Delta W/g_J\hbar$, allowing universal representations of the Zeeman effect of the HFS which depend only on the spin structure. Figure 8.3 shows these level diagrams for $J = 1/2$ $I = 1/2$ such as for H and $J = 1/2$; $I = 1$, such as for D.

Determination of the *eigenvectors* (i.e. wave functions of the different HFS states): *Definition*: $(H - \lambda_{\pm})\psi_{\pm} = 0$. The functions ψ, according to definition are linear combinations of the base vectors $| \pm \frac{1}{2}, m_F \mp \frac{1}{2}\rangle$:

$$|\psi_{\pm}\rangle = \alpha_{\pm}|\frac{1}{2}, m_F - \frac{1}{2}\rangle + \beta_{\pm}|-\frac{1}{2}, m_F + \frac{1}{2}\rangle \qquad (8.21)$$

With the normalization condition $|\alpha_{\pm}|^2 + |\beta_{\pm}|^2 = 1$ and the orthogonality condition $\langle \psi_+|\psi_-\rangle = 0$ the solutions are obtained from:

$$\begin{pmatrix} H_{11} - \lambda_{\pm} & H_{12} \\ H_{12} & H_{22} - \lambda_{\pm} \end{pmatrix}\begin{pmatrix} \alpha_{\pm} \\ \beta_{\pm} \end{pmatrix} = \begin{pmatrix} 0 \\ 0 \end{pmatrix}. \qquad (8.22)$$

Example: for the third Zeeman state of deuterium one obtains

$$\lambda_+ = E_3 = -\frac{\Delta W}{2}\left(\frac{1}{3} - \sqrt{1 - \frac{2}{3}x + x^2}\right) \quad \text{follows} \qquad (8.23)$$

$$|\psi_3\rangle = |\psi_+\rangle = c|\frac{1}{2}, -1\rangle + d|-\frac{1}{2}, 0\rangle \quad \text{with} \qquad (8.24)$$

$$c = \frac{1}{\sqrt{2}} \left[1 + \left(x - \frac{1}{3} \right) \Big/ \sqrt{1 - \frac{2}{3}x + x^2} \right]^{1/2} \quad \text{and} \quad (8.25)$$

$$d = \frac{1}{\sqrt{2}} \left[1 - \left(x - \frac{1}{3} \right) \Big/ \sqrt{1 - \frac{2}{3}x + x^2} \right]^{1/2}, \quad (8.26)$$

and similar for all other states.

8.2.3 Zeeman Splitting of the H Atom

Again numbering the four Zeeman states from up to down one obtains for the eigenvalues

$$E_1 = \frac{\Delta W}{2} \left[-\frac{1}{2} + (1 + x) \right] \quad (8.27)$$

$$E_2 = \frac{\Delta W}{2} \left[-\frac{1}{2} + \sqrt{1 + x^2} \right] \quad (8.28)$$

$$E_3 = \frac{\Delta W}{2} \left[-\frac{1}{2} + (1 - x) \right] \quad (8.29)$$

$$E_4 = \frac{\Delta W}{2} \left[-\frac{1}{2} - \sqrt{1 + x^2} \right] \quad (8.30)$$

and the wave function with

$$a' = \frac{x}{\sqrt{1 + x^2}} \quad (8.31)$$

$$|1\rangle = \left| \frac{1}{2}, \frac{1}{2} \right\rangle \quad (8.32)$$

$$|2\rangle = \frac{1}{\sqrt{2}} \left(\sqrt{1 + a'} \left| \frac{1}{2}, -\frac{1}{2} \right\rangle + \sqrt{1 - a'} \left| -\frac{1}{2}, \frac{1}{2} \right\rangle \right) \quad (8.33)$$

$$|3\rangle = \left| -\frac{1}{2}, -\frac{1}{2} \right\rangle \quad (8.34)$$

$$|4\rangle = \frac{1}{\sqrt{2}} \left(\sqrt{1-a'} \left| \frac{1}{2}, -\frac{1}{2} \right\rangle - \sqrt{1+a'} \left| -\frac{1}{2}, \frac{1}{2} \right\rangle \right) \qquad (8.35)$$

Again $x = B/B_{crit}$, and for the three relevant states B_{crit} and ΔW have the values:

State	B_{crit}[mT]	ΔW [MHz]
$1S_{1/2}$	50.7	1420
$2P_{1/2}$	2.1	60
$2S_{1/2}$	6.4	178

8.2.4 Zeeman Splitting of the D Atom

The eigenvalues of the six states (again numbered as above) are:

$$E_1 = \frac{\Delta W}{2} \left[-\frac{1}{3} + (1+x) \right] \qquad (8.36)$$

$$E_2 = \frac{\Delta W}{2} \left[-\frac{1}{3} + \sqrt{1 + \frac{2}{3}x + x^2} \right] \qquad (8.37)$$

$$E_3 = \frac{\Delta W}{2} \left[-\frac{1}{3} + \sqrt{1 - \frac{2}{3}x + x^2} \right] \qquad (8.38)$$

$$E_4 = \frac{\Delta W}{2} \left[-\frac{1}{3} + (1-x) \right] \qquad (8.39)$$

$$E_5 = \frac{\Delta W}{2} \left[-\frac{1}{3} - \sqrt{1 - \frac{2}{3}x + x^2} \right] \qquad (8.40)$$

$$E_6 = \frac{\Delta W}{2} \left[-\frac{1}{3} - \sqrt{1 + \frac{2}{3}x + x^2} \right] \qquad (8.41)$$

and the corresponding wave functions with

$$a = \frac{x + \frac{1}{3}}{\sqrt{1 + \frac{2}{3}x + x^2}} \quad \text{and} \quad b = \frac{x - \frac{1}{3}}{\sqrt{1 - \frac{2}{3}x + x^2}} \qquad (8.42)$$

$$|1\rangle = \left|\frac{1}{2}, 1\right\rangle \tag{8.43}$$

$$|2\rangle = \frac{1}{\sqrt{2}}\left(\sqrt{1+a}\ \left|\frac{1}{2}, 0\right\rangle + \sqrt{1-a}\ \left|-\frac{1}{2}, 1\right\rangle\right) \tag{8.44}$$

$$|3\rangle = \frac{1}{\sqrt{2}}\left(\sqrt{1+b}\ \left|\frac{1}{2}, -1\right\rangle + \sqrt{1-b}\ \left|-\frac{1}{2}, 0\right\rangle\right) \tag{8.45}$$

$$|4\rangle = \left|-\frac{1}{2}, -1\right\rangle \tag{8.46}$$

$$|5\rangle = \frac{1}{\sqrt{2}}\left(\sqrt{1-b}\ \left|\frac{1}{2}, -1\right\rangle - \sqrt{1+b}\ \left|-\frac{1}{2}, 0\right\rangle\right) \tag{8.47}$$

$$|6\rangle = \frac{1}{\sqrt{2}}\left(\sqrt{1+a}\ \left|-\frac{1}{2}, 1\right\rangle - \sqrt{1-a}\ \left|\frac{1}{2}, 0\right\rangle\right) \tag{8.48}$$

State	B_{crit} (mT)	ΔW (MHz)
$1S_{1/2}$	11.7	327
$2P_{1/2}$	0.5	14
$2S_{1/2}$	1.5	41

8.2.5 Calculation of Polarization

For the calculation of the (electronic or nuclear) polarization of a system with coupled nuclear and electronic spins it is useful to calculate the polarization of each Zeeman component and to take the weighted average over all *occupied* components, as shown in Fig. 8.4. The (field dependent) occupation numbers of the Zeeman states for the calculation of the polarization are the squares of the amplitudes of these states. One uses the common definitions of the polarization

$$p^* = \frac{N_+ - N_-}{N_+ + N_-} \quad \text{for H   and} \tag{8.49}$$

$$p^* = \frac{N_+ - N_-}{N_+ + N_- + N_0} \tag{8.50}$$

$$p^*_{zz} = \frac{N_+ + N_- - 2N_o}{N_+ + N_- + N_o} \quad \text{for D} \tag{8.51}$$

The symbol * signifies the polarization with respect to quantization axis of the states (in general the magnetic field direction at the ionizer). It is a quantity independent of the coordinate system, and also part of a "figure of merit" which measures e.g. the "quality" of an ensemble of polarized particles such as a source or target (the true figure of merit is—for statistics reasons, i.e. the quantity which should be maximized for minimum measurement time for given precision—the intensity or density times the square of the polarization). It is identical with the z (or zz) component of polarization considered, if the z axis coincides with the quantization axis and is equal to the absolute maximum polarization. Therefore, these quantities are sometimes written with capital-letter indices as p_Z, p_{ZZ}. The density matrix for this case appears in diagonal form. Under rotations the components of the "polarization" behave according to their tensor character with the appropriate rotation functions (spherical harmonics, Legendre polynomials or functions), e.g. the vector polarization $\vec{p}$ as a first-degree tensor with $P_1(\cos\beta)$ or $P_1^1(\cos\beta)$.

An example For H we assume that the states $|1\rangle$ and $|2\rangle$ are occupied, e.g. behind a Stern–Gerlach magnet. Then the nuclear-spin occupation numbers are:

$$N_{+\frac{1}{2}}(I) = \frac{1}{2}\left[1 + \frac{1}{2}(1 - a')\right] = \frac{1}{2}\left(\frac{3}{2} - \frac{1}{2}\frac{x}{\sqrt{1+x^2}}\right) \tag{8.52}$$

$$N_{-\frac{1}{2}}(I) = \frac{1}{2}\left[\frac{1}{2}(1 + a')\right] = \frac{1}{4}\left(1 + \frac{x}{\sqrt{1+x^2}}\right) \tag{8.53}$$

Thus the limiting cases are obtained:

$$for \quad x \to 0 : N_{+\frac{1}{2}} = \frac{3}{4} \tag{8.54}$$

$$N_{-\frac{1}{2}} = \frac{1}{4} \qquad p^*(I) = \frac{1}{2} \tag{8.55}$$

$$for \quad x \to \infty : N_{+\frac{1}{2}} = \frac{1}{2} \tag{8.56}$$

$$N_{-\frac{1}{2}} = \frac{1}{2} \qquad p^*(I) = 0 \tag{8.57}$$

For the electronic polarization the same arguing gives:

$$N_{+\frac{1}{2}}(J) = \frac{1}{2}\left[1 + \frac{1}{2}(1 + a')\right] = \frac{1}{2}\left(\frac{3}{2} + \frac{1}{2}\frac{x}{\sqrt{1+x^2}}\right) \tag{8.58}$$

$$N_{-\frac{1}{2}}(J) = \frac{1}{2}\left[\frac{1}{2}(1 - a')\right] = \frac{1}{4}\left(1 - \frac{x}{\sqrt{1+x^2}}\right) \tag{8.59}$$

and thus for $x \to 0$: $p^*(J) = \frac{1}{2}$ and for $x \to \infty$: $p^*(J) = 1$.

Fig. 8.4 Polarization of each single Zeeman component of hydrogen and deuterium. The states are numbered as usual, see Fig. 8.3. The polarization of an ensemble of particles (e.g. in a beam), which are in different Zeeman states is obtained by performing the weighted average over all occupied components

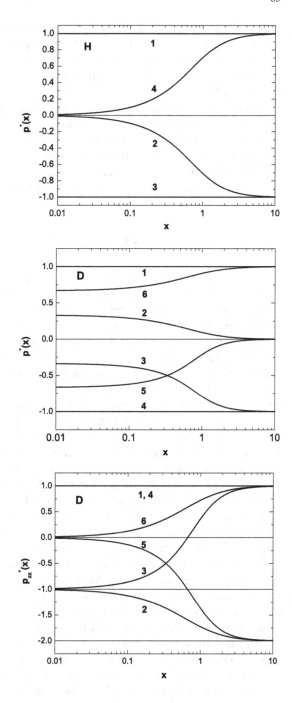

During an *adiabatic* transition from a very strong into a very weak magnetic field the complete electronic polarization in a strong Stern–Gerlach field is distributed over

the previously unpolarized nuclei by the hyperfine interaction such that subsequently both are partly polarized (for H: each half-polarized).

("Adiabatic" means: the relative change of the magnetic field per unit time has to be small as compared with the Larmor frequency: $dB/(Bdt) \ll \omega_L = \gamma B$, i.e. $dB/dt \ll \gamma B^2$).

Further examples of the calculation of the beam polarization (from a source) in different modes of operation are given below.

Stern–Gerlach separation of HFS states in a ground-state atomic beam source: the separation strength of the HFS states is given by the force of the inhomogeneous magnetic field acting on the "effective magnetic moment". This is defined as the derivative of the energy $W(B)$, given by the Breit-Rabi formula (see below), after the field strength B:

$$\vec{F} = -\vec{\nabla} W_{F,m_F} = -\frac{\partial W}{\partial B} \vec{\nabla} B = \mu_{eff} \vec{\nabla} B \qquad (8.60)$$

Only for the "pure" components is $\mu_{eff} = \mu_B$. Therefore, the separation according to the m_J works only for large B. Figure 8.5 shows the effective magnetic moments of the Zeeman components as functions of the magnetic field. In order to increase the beam intensity by a separation of the Zeeman components in two dimensions multipole fields with cylindrical symmetry (quadrupole or sextupole fields) are used in practice. The principle of constructing a polarized-ion source, e.g. for use on accelerators and thus obtaining much higher intensities at high beam quality and complete control over the polarization parameters, as compared to using nuclear reactions as primary source of polarized particles, was first formulated by Clausnitzer et al. [10, 11]. The first nuclear reaction initiated by a polarized beam from such a source was the ^{3}H($\vec{d}$, n) ^{4}He reaction on resonance at $E_d = 107$ keV at Basel [12] which was the occasion for the first polarization conference [13].

The radial dependence of the force on magnetic moments is given by r^{L-2} (L = multipole order of the magnetic field). Thus, the force in a quadrupole field is constant and that in a sextupole is linear in r. In a sextupole there is a lens-like focussing action on one spin component whereas the other is being defocussed. Therefore, an "optics" for spin-magnetic moments with features like beam transport, phase space, emittance, acceptance in analogy to the optics of charged particles in electric fields can be defined and be used to optimize a Stern–Gerlach system ("matching"). Note, however, that the sextupole provides no state separation on the axis, whereas for the quadrupole the state separation force is uniform with r, but there is no focussing. For several reasons (among these better pumping, the requirement of leaving space for intermediate radiofrequency transitions, and higher flexibility to optimize the atomic-beam optics) modern ground-state atomic-beam polarized ion sources (ABS) use not one, but a number of spin-separation magnets. It is suggestive to use a quadrupole magnet as first magnet leading to a better spin-state separation and somewhat higher polarization. The main quantities characterizing an ABS are the polarization p, the beam intensity I, but also the beam quality ("brightness" = intensity per transverse emittance). From the point of view of minimizing the

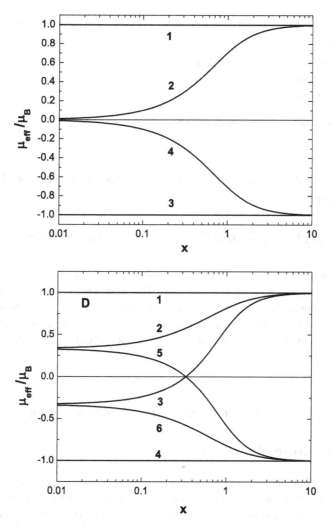

Fig. 8.5 "Effective" magnetic moments in units of μ_B of the HF components of H and D in an inhomogeneous magnetic field with $x = B/B_{crit}$. The numbering is as in Fig. 8.3. The deflecting force in an inhomogeneous magnetic field B is $\mu_{eff} \cdot \nabla |\vec{B}|$

measurement time for a given statistical error in experiments the figure of merit is $p^2 I$ which is valid for vector and tensor polarization components. When ionization of the neutral beam takes place in a strong magnetic field the ion beam acquires transverse momentum thus increasing the transverse phase space, i.e. the emittance. Therefore the usual atomic-beam sources (ABS) have emittances (typically 2 cm rad $(eV)^{1/2}$) about twice those of Lambshift (LSS) and colliding-beams (CBS) sources (typically <1 cm rad $(eV)^{1/2}$).

With the above outlines the principles of common types of polarized-ion sources can be understood. These are:

- Ground-state atomic-beam sources (ABS). They differ in the way the atomic beam is ionized:

  - Electron-bombardment and ECR ionizers
  - Ionizers with colliding beams of Cs^0, H, or D
  - Optically-pumped ion sources

- Lambshift polarized-ion sources (LSS)

8.3 Physics and Techniques of the Ground-State Atomic Beam Sources ABS

8.3.1 Production of H and D Ground-State Atomic Beams

In order to produce atomic beams of H/D dissociators of different designs, all based on radiofrequency (RF) excitation, are used. The atomic beam intensity depends on a number of parameters: gas pressure and gas flux, RF power, recombination rate on surfaces and their temperatures, and intra-beam scattering processes. After many years of development (since 1965) optimal design schemes have evolved which will be described here.

8.3.2 Dissociators, Beam Formation and Accomodation

8.3.2.1 RF-Discharge Dissociators

Two types of dissociators have evolved. In both a gas discharge excited by an RF field are used. The classical method has been to use a cylindrical Pyrex glass or quartz vessel with H_2 or D_2 being fed in from one end and H or D atoms streaming out at the other. The discharge is maintained by a coil around the glass bottle (magnetic coupling) and normally runs at about 13 MHz at an RF power up to 200 W. The proper matching of the discharge assembly to a power oscillator is achieved by a matching circuit. The atoms are at or slightly above room temperature and are formed into a beam by a nozzle, typically from aluminum with a (sometimes tapered) canal 8–20 mm long and an orifice 2–3 mm wide. Their velocity distribution is not purely Maxwellian but somewhat narrower due to the action of the nozzle. The nozzle is cooled to either in the range 50–100 K or, in a different mode, to about 30 K. The cooling is essential for several reasons. The intensity of the beam is determined by a number of parameters: The discharge is burning best in a certain pressure range within which the gas feed should be as high as possible. The limit is, however, set by the pumping speed with which the space after the nozzle can be maintained at such a low pressure that the mean free path is long enough to

Fig. 8.6 Typical
nozzle-skimmer-collimator
arrangement of ABS sources.
Typical diameters: nozzle
1 mm, skimmer 1.5 mm

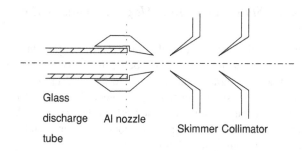

Glass

discharge   Al nozzle
                          Skimmer Collimator
tube

avoid or at least minimize intra-beam scattering of the atoms. These conditions
are constrained by the small available space for pumping thus limiting the conductance, i.e. the effective pumping speed in this space. Differential pumping has to
be applied and better beam quality is achieved by a skimmer. The collimator is
necessary for differential pumping and facilitates maintaining high vacuum in the
separation-magnet regions. A typical setup is shown in Fig. 8.6. As nozzle materials
copper and aluminum have been used because of their heat conductivity but due
to layers formed on the inner surface the choice is not critical for recombination.
For the reasons discussed above, over the years a saturation of the polarized atomic-
beam intensities at slightly above 10^{17} atoms/s in the relevant region behind the last
separation magnet is observed.

The role of cooling the dissociator arrangement is twofold: first the dissociator
vessel must be cooled to prevent the glass from heating up (causing background
residual gas and increasing recombination), secondly the nozzle must be cooled to
make the atoms slower. The effect of this is that the acceptance of the entrance
to the separation magnet system is increased, the separation power of the magnets
improves, and the ionization yield of an ionizer increases because the beam density
$\rho = j/v$ is higher (j = beam particle current density, v = particle velocity). Some
modelling showed that the sum of these effects scales as $\propto T^{-3/2}$ [14]. However,
because the velocity distribution is not Maxwellian (the real distribution is narrower
and displaced [15]) the effects are more complicated in detail, see e.g. Singy et al.
[14]. It was shown that cooling the nozzle much below 80 K resulted in increased
recombination on the nozzle surface. Especially below about 45 K the recombination
increases sharply resulting in a severe drop of output beam intensity. This could be
partly remedied by adding some N_2 to the gas. Additional adding of some O_2 seems
to improve the degree of dissociation.

8.3.2.2 Microwave Dissociators

At the HERMES/DESY PIS a microwave dissociator has been developed.
At microwave power levels of 400–1000 W and RF frequency of 2.45 GHz higher
gas flow was possible [16].

8.3.3 State-Separation Magnets: Classical and Modern Designs

Historically starting from the Stern–Gerlach spin-state separation magnet working in one dimension only much better intensity can be achieved with rotationally-symmetric magnetic fields provided by quadrupole and sextupole magnets acting in two dimensions. Permanent magnets as well as electromagnets have been used where the latter could be turned off for an unpolarized beam. However, the advantages of permanent magnets of modern design ("Halbach" magnets) are such that almost all sources use them.

8.3.3.1 Multipole Fields

The properties of magnetic multipole fields are:

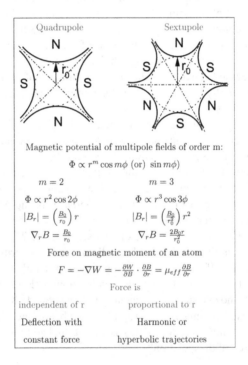

Magnetic potential of multipole fields of order m:

$$\Phi \propto r^m \cos m\phi \; (\text{or}) \; \sin m\phi$$

$m = 2$	$m = 3$
$\Phi \propto r^2 \cos 2\phi$	$\Phi \propto r^3 \cos 3\phi$
$|B_r| = \left(\frac{B_0}{r_0}\right) r$	$|B_r| = \left(\frac{B_0}{r_0^2}\right) r^2$
$\nabla_r B = \frac{B_0}{r_0}$	$\nabla_r B = \frac{2B_0 r}{r_0^2}$

Force on magnetic moment of an atom

$$F = -\nabla W = -\frac{\partial W}{\partial B} \cdot \frac{\partial B}{\partial r} = \mu_{eff} \frac{\partial B}{\partial r}$$

Force is

independent of r	proportional to r
Deflection with	Harmonic or
constant force	hyperbolic trajectories

Only sextupole fields focus the atoms like an optical lens. Atoms in the opposite spin states are defocused. It is obvious that on the axis (in an infinitesimal volume) in a sextupole there is no spin-state separation whereas in a quadrupole the separation force is constant over the entire volume. At least in principle this should guarantee a somewhat higher beam polarization than from a sextupole. The rather wide velocity distribution of the atoms leads to a strong chromatic aberration of multipole magnets. This can be partly offset by tapering the magnetic fields along the z axis.

Fig. 8.7 Cross-section of a "classical" permanent sextupole magnet of length 25 cm (*left*) and MoO_3 picture of a hydrogen beam behind this magnet showing a concentration of the $m_J = +1/2$ atoms in the center (*right*)

Originally the multipole magnets—only permanent-magnets designs will be discussed—consisted of separate polepieces formed by permanent magnets with high remanence (such as ALNICO V) ending in angular poletips from magnetically soft materials (such as soft iron or Vanadium-Permendur). A soft-iron cylinder provided the closing of the magnetic circuit. Typical pole-tip fields were about 1 T. Figure 8.7 shows a cross-section view of such a magnet which was part of a very compact polarized-ion source specially designed to work under the extreme conditions inside the high-voltage terminal of a single-ended CN Van-de-Graaff accelerator, and the concentration of $m_J = +1/2$ atoms in the center [17, 18]. The focussing action of this magnet on a beam of hydrogen atoms was shown using the old technique of MoO_3, a yellowish material being reduced to dark Mo by the atomic hydrogen, thereby producing an image of the atom distribution.

About 1980 a new design of multipole magnets [19] emerged with a number of improvements. Besides new and stronger magnetic materials which could be magnetized in arbitrary directions the magnets have much less surface area with less outgassing and are much smaller in size. The Halbach 2N-pole magnets consist of M segments magnetized in changing directions such that after M/2N segments there is a pole-piece segment with opposite orientation (e.g. from a north to a south pole) and the segments between them have skewed magnetization directions. Figure 8.8 [20] shows the setup of a typical Halbach sextupole and a measured field distribution.

Recently a modification of the Halbach design was suggested [21] for a quadrupole magnet and first realized as a sextupole at Cologne [22]. In this design the high-field regions of the polepieces are replaced by soft iron material thus concentrating the field lines. The first such sextupole has been produced in 2003 by "Vacuumschmelze"[3] and successfully used in the Cologne SAPIS source project [23, 24]. The measured maximum field values did not quite fulfill the theoretical expectations but nevertheless 1.6 T were obtained with potential for improvement. Figure 8.9 shows the Cologne

---

[3] VACUUMSCHMELZE GmbH & Co. KG, D-63412 Hanau. Materials used: VACODYM 633 HR, 677 HR, and VACOFLUX 50.

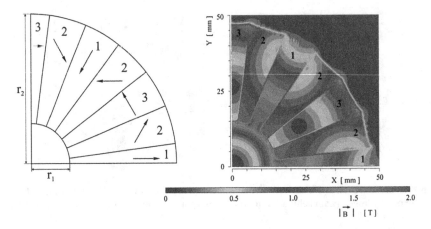

Fig. 8.8 Segment scheme and field distribution of a typical Halbach sextupole magnet [20]

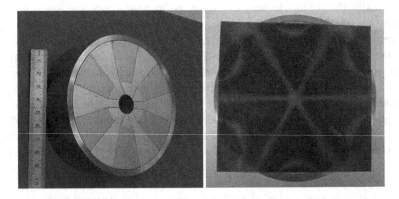

Fig. 8.9 View of the modified Halbach ("hybrid") sextupole magnet developed for SAPIS-Cologne with a measured maximum field value of 1.6 T. On the right a field image is shown obtained with "Magna View Film"

sextupole magnet together with a field image on "Magna View" film.[4] In the meantime other applications have emerged, e.g. for focussing cold pulsed neutrons [26].

In modern ABS an arrangement of several (typically four to six) separated short sextupoles is used. The field strengths and location of the magnets are determined by numerical trajectory calculations taking into account other requirements such as optimum pumping and insertion of RF transition units. Figure 8.10 shows one such calculation.

---

[4] From Edmund Scientific, http://scientificsonline.com (2004).

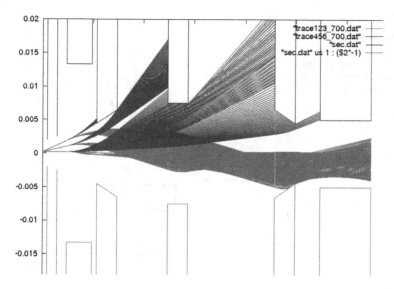

Fig. 8.10 Result of on optimized trajectory calculation for a deuterium beam through one quadrupole and three sextupole magnets

8.3.4 RF Transitions

After a Stern–Gerlach device and in a strong magnetic field (e.g. in an electron-bombardment ionizer) the particles in a beam are highly polarized with respect to the spin of their valence electron, but nearly unpolarized in nuclear spin. By guiding the atoms adiabatically into a region of a weak magnetic field nuclear polarizations with one half (H) or one third (D) of the maximum values are achieved. For higher nuclear polarization the occupation numbers of the hyperfine states of these slow (sub-thermal to thermal) atomic beams have to be changed. Also, for higher intensities ionization in a strong magnetic field is necessary (see below). The large number and type of possible RF transitions and the combination of several transition units allow all kinds of different (vector and or tensor) polarizations in different combinations and sign changes. Thus one can get all the flexibility of having maximum polarization values, purely vector and/or tensor polarized beams and the possibility of changing signs of the polarizations without changing the beam geometries, thus avoiding instrumental asymmetries.

For the necessary RF transitions between different Zeeman hyperfine states the adiabatic-fast passage method was proposed by Abragam and Winter [25]. It is used in all ABS and polarized-beam targets and provides high polarization (near the theoretical maximum). Due to the field gradient the transitions are independent of the velocity distribution, the resonances are easy to find and allow stable operation. A semi-classical picture can be used to illustrate the transition while the beam passes through a magnetic field varying linearly from $B_0 - \Delta B$ to $B_0 + \Delta B$ (or in reversed

Fig. 8.11 Semi-classical explanation of the "adiabatic-fast passage" method of producing a spin flip during passage through an inhomogeneous magnetic field

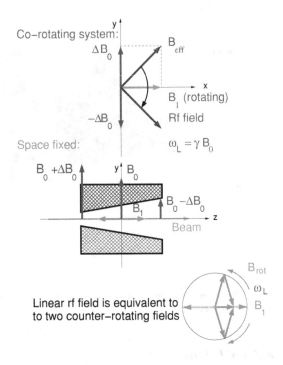

order) with the resonance at B_0 and the corresponding RF frequency $\nu = \mu B/h$ for a complete spin reversal for atoms with velocity v. The spin I with magnetic moment μ_I precesses with the corresponding angular velocity $\omega = 2\pi \nu$. As best seen by a co-rotating observer a rotating RF field B_1 exerts a torque on μ such that during passage through the gradient field the spin just moves from up to down with (theoretically) 100% efficiency. The rotating field can be imagined as one of two counter-rotation fields (the other has no effect) equivalent to a linear RF field. This is illustrated in Fig. 8.11. RF hyperfine transitions may be approximately classified according to the value of the static magnetic field B_0, e.g. in relation to the critical field B_{crit} : weak-field (WFT, $B_0 \ll B_{crit}$, transition frequencies typically 5–15 MHz), medium-field (MFT, $B_0 < B_{crit}$), and strong-field (SFT, $B_0 \geq B_{crit}$, transition frequencies typically several hundred MHz–GHz) transitions. Another classification refers to the change of quantum numbers by the transitions. Transitions within one F multiplet ($\Delta F = 0$, $\Delta m_F = \pm 1$) are π transitions and they are induced by the RF field $B_1 \perp B_0$. Transitions between different F multiplets ($\Delta F = \pm 1$, $\Delta m_F = 0, \pm 1$) are σ transitions, and the two fields are parallel to each other, see e.g. Ramsey [27].

In the practice of polarized-ion sources the WFT and MFT used are low-B_0 π transitions. The WFT occur in the Zeeman region of the HFS where the m_F states belonging to one F are nearly equidistant, leading to multi-quantum transitions within the F multiplets. The MFT are similar π transitions at somewhat higher B_0 and RF frequencies such that the energies of single-photon transitions in one F multiplet are sufficiently separated, i.e. with a field region short enough that only single transi-

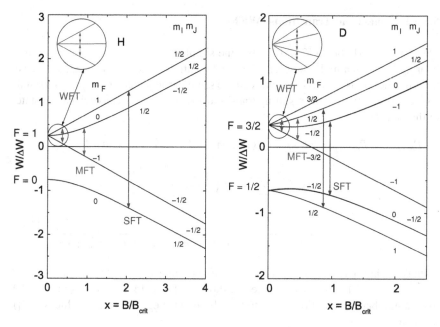

Fig. 8.12 RF transitions and transition types as functions of the field parameter x

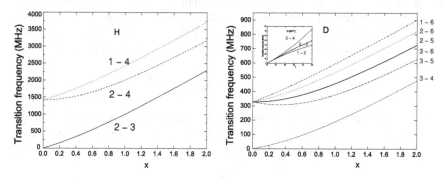

Fig. 8.13 Transition frequencies of selected RF transitions as functions of the field parameter x

tions do occur. SFT, however, are σ or π single-quantum transitions at still higher B_0 between single HFS states. Figure 8.12 illustrates the three types of transitions whereas Fig. 8.13 depicts the transition frequencies for a number of rf transitions as functions of the magnetic field parameter x.

8.3.4.1 Weak-Field Transitions (WFT)

The theory of Majorana [28] describes the simultaneous transition between neighbouring levels in a multiplet ($\Delta F = 0$, π transitions) with $\Delta m_F = \pm 1$ by transitions in a fictitious spin-1/2 system with the same gyromagnetic ratio γ. His formula
e.g. Ramsey [27] gives the resulting transition probability between each pair of states
of the multiplet:

$$
P_{m_F m'_F} = P_{m'_F m_F} = (F - m_F)!(F + m_F)!(F - m'_F)!(F + m'_F)!
$$

$$
\cdot p^{2F} \left[\sum_n (-1)^n \left(\frac{1 - p}{p} \right)^{\frac{m_F + m'_F}{2} + n} \right.
$$

$$
\left. \cdot \frac{1}{(F - m_F - n)!(F - m'_F - n)!(m_F + m'_F + n)!n!} \right]^2 .
\tag{8.61}
$$

With this relation the occupation numbers of hyperfine states and the polarization of
a beam after a WFT can be calculated as shown in Fig. 8.14. For the design of WFT
units a number of conditions have to be fulfilled (see eg. Paetz gen. Schieck [29]).
These are

1. Equidistance condition: For the Majorana approach to hold the assumption of
   nearly equidistant level separation should be fulfilled. By expanding the Breit-
   Rabi equations of the HFS Zeeman states of H or D one obtains the deviation
   from a linear splitting of neighboring levels with energies E_n, E_{n+1}, and E_{n+2}

$$
\epsilon = (E_n - E_{n+1}) - (E_{n+1} - E_{n+2}) \approx 2(b_F \frac{\mu_B}{\hbar} B)^2 \frac{\hbar^2}{\Delta W}
\tag{8.62}
$$

   Classically the perturbation by the energy injected into the system by the RF field
   B_1 should be much larger than ϵ in order to effectuate the transition. Therefore,
   $\hbar \mu_B b_F B_1 \gg \epsilon$ which translates into

$$
B_1 \gg \frac{2 b_F \mu_B B_0^2(0)}{\Delta W} .
\tag{8.63}
$$

2. Field non-reversal condition: The magnetic field must not undergo a sign change
   which leads to

$$
\Delta B_0 < B_0(0).
\tag{8.64}
$$

3. Adiabaticity condition: The transition must be adiabatic. In a semi-classical
   picture for a complete transition the spin must follow the magnetic field over
   the entire transition which requires the relative time change of the field B_0 to be
   slow as compared to the inverse of the Larmor frequency. The transit time through
   the transition region may be expressed by the length ℓ and velocity $\bar{v}$. Thus the
   conditions read

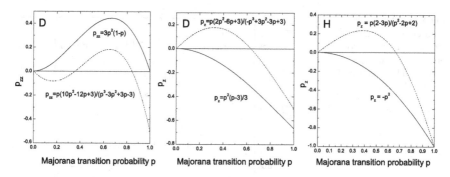

Fig. 8.14 Nuclear polarization of H and D in a strong magnetic field as a function of the WFT transition probability p. Plotted is the polarization after the first Stern–Gerlach device (states 1 and 2/ 1,2, and 3 occupied) and subsequent WFT (*solid line*), in addition after another Stern–Gerlach magnet with 100% spin-state separation efficiency (*dashed line*)

$$|\dot{B_0}| \ll \frac{b_F \mu_B B_1^2}{2\hbar}, \tag{8.65}$$

or equivalently:

$$B_1^2 \gg \frac{4\bar{v}\hbar}{b_F \ell \mu_B} \Delta B_0. \tag{8.66}$$

4. Polarization condition: In order to achieve a complete spin flip of (classically) 180° (for a semi-classical explanation see Haeberli [30]) the condition is

$$\frac{B_1}{2} \ll \Delta B_0. \tag{8.67}$$

5. Resonance condition: Since the transition is a resonant phenomenon this condition is trivial

$$\omega_0 = \frac{2\mu_B b_F B_0}{\hbar}. \tag{8.68}$$

Conditions 1–4 define an allowed region of the necessary parameters for the proper choice of a working point as shown in Fig. 8.15.

8.3.4.2 Quantum-Mechanical Treatment

The quantum-mechanical transformation of the two-spin system with time-varying magnetic fields (the "static" field B_0 slowly varying due to the particle motion in its field gradient $2\Delta B_0/\Delta z$ and the (fast-changing) RF field $B_1(\omega t)$) into a co-rotating system is equivalent to unitary transformations of the Hamiltonian of the system.

Fig. 8.15 Plot of the
semi-classical criteria for the
choice of the WFT working
point. The numbering
corresponds to the list of
criteria above

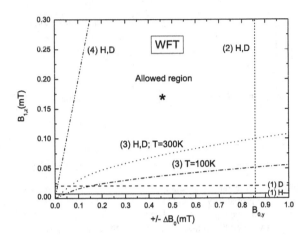

If the transformation is appropriately chosen (i.e. such that the rotation occurs with
the Larmor frequency ω) the slowly- and the fast-varying parts can be separately
diagonalized. The "quasi-stationary" slow-solution part of the Schrödinger equation
leads just to a different picture of the Breit-Rabi energy eigenstates (as functions
of B_0 or, equivalently, x, or t) which follow a linear dependence making the states
cross at just the field corresponding to the particular transition frequency. With the
RF field B_1 switched on, the q.m. calculation including this perturbation leads to (in
this representation)

- up and down shifts of the eigenvalues ("level repulsion") and to
- non-diagonal terms of the Hamiltonian matrix, i.e. mixing of states and therefore
  transitions between them.

Figure 8.16 shows this schematically. The effect is strongest at the crossing
points, and the efficiency of the transition is governed by the degree of adiabaticity,
i.e. whether the occupation of the initial states involved will stay on its original levels
(non-adiabatic or "diabatic" transitions, see also Sect. 8.5.3) or undergo a more or
less complete transition to the other level. The strength of the RF field B_1 is one
determining factor. In order to quantify this degree an adiabaticity parameter was
derived from the adiabaticity condition Eq. (8.66)

$$\kappa = \frac{\mu_J B_1^2}{2\hbar \dot{B}_0} \tag{8.69}$$

leading to the transition probability

$$P = \exp(-\pi \kappa) \tag{8.70}$$

(for details see Philpott [31], Beijers [32]). The formalism is applicable to two-state
strong-field as well as medium-field transitions.

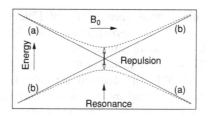

Fig. 8.16 Schematical q.m. representation of two Zeeman hyperfine states (a) and (b) as functions of the external static magnetic gradient field B_0 without perturbing RF field B_1 : states crossing (*solid lines*) and with perturbation: mixing, repulsion of states, and more or less transition between them (*dashed lines*)

Detailed information about the RF transitions is obtained by the solution of the time-dependent Schrödinger equation, in the case of multiple transitions like the WFT by a set of $2m_F + 1$ coupled, time-dependent equations. The equations are

- For the F = 1 triplet of hydrogen we need three equations

$$\dot{c}_1(t) = -2i\Omega(t)\cos(\omega t)e^{i\omega_{12}(t)t/\hbar}c_2(t)$$
$$\dot{c}_2(t) = -2i\Omega(t)\cos(\omega t)e^{-i\omega_{12}(t)t/\hbar}c_1(t) - 2i\Omega(t)\cos(\omega t)e^{i\omega_{23}(t)t/\hbar}c_3(t)$$
$$\dot{c}_3(t) = -2i\Omega(t)\cos(\omega t)e^{-i\omega_{23}(t)t/\hbar}c_3(t)$$

(8.71)

- For the F = 3/2 multiplet of deuterium—the states with F = 1/2 can be neglected— we have four equations

$$\dot{c}_1(t) = -2i\Omega(t)\cos(\omega t)e^{i\omega_{12}(t)t/\hbar}c_2(t)$$
$$\dot{c}_2(t) = -2i\Omega(t)\cos(\omega t)e^{-i\omega_{12}(t)t/\hbar}c_1(t) - 2i\Omega(t)\cos(\omega t)e^{i\omega_{23}(t)t/\hbar}c_3(t)$$
$$\dot{c}_3(t) = -2i\Omega(t)\cos(\omega t)e^{-i\omega_{23}(t)t/\hbar}c_2(t) - 2i\Omega(t)\cos(\omega t)e^{i\omega_{34}(t)t/\hbar}c_4(t)$$
$$\dot{c}_4(t) = -2i\Omega(t)\cos(\omega t)e^{-i\omega_{34}(t)t/\hbar}c_3(t)$$

(8.72)

The parameters of these equations, which are partly time-dependent due to the beam moving through varying magnetic fields, are: ω the externally applied circular RF frequency, $\omega_{nm}(t = z/\bar{v})$ the circular Bohr transition frequencies between neighbouring hyperfine states n and m (see Fig. 8.13), and $\Omega(t = z/\bar{v}) = \mu_B B_1(t)/2\hbar$ the "Bloch" circular frequency which depends on the amplitude B_1 of the external RF field and is a measure of the transition strength.

Experimentally as well as in theoretical studies differences in occupation numbers and therefore polarizations have been found depending on the sign of the gradient of the static field B. It was argued that the equidistance condition was not perfectly fulfilled and that the several transitions between Zeeman-HFS states did not occur simultaneously but sequentially. The sequence depended on the sign of the gradient ΔB_0, see Philpott [31], Glavish [33].

Several such calculations have been published [29, 32, 34] and the final occupation numbers of the Zeeman hyperfine states and thus the nuclear or electronic polarizations following the transitions have been evaluated as functions of a number of parameters. These were e.g. the particle velocities and the strength of the inducing RF field B_1. For the WFT the observed fact that in some cases the polarization depended on the sign of the gradient ΔB_0 was confirmed when the amplitude of the B_1 field was not sufficiently high. The explanation lies in the fact that the transitions take place at some small field $B_0 > 0$ where the states of the m_F are not completely equidistant and the single transitions occur in different sequences for the two cases except for a sufficiently high amplitude of B_1. This is in qualitative agreement with the results using the Majorana formalism, see Fig. 8.14. Figure 8.17 gives one example of a complete (B_1 high enough) and an incomplete (B_1 too small) spin flip for the state 1 of deuterium. The same can be calculated for all relevant states (for details see Paetz gen. Schieck [29]). For the WFT being π transitions the direction of the RF field B_1 is along the beam axis (z direction) and perpendicular to the static field B_0. Due to the low frequency required it is realized by a coil with a small number of windings (e.g. about 5–10 for frequencies of 8–12 MHz). Typical values are given in Paetz gen. Schieck [29].

8.3.4.3 Medium-Field Transitions (MFT)

Like the SFT the MFT are transitions between single states. They are, however, π transitions occuring at rather low B_0. Typical transitions are between states $1 \leftrightarrow 2$ and $2 \leftrightarrow 3$ for H and $1 \leftrightarrow 2$, $2 \leftrightarrow 3$, $3 \leftrightarrow 4$, and $5 \leftrightarrow 6$ for D.

8.3.4.4 Strong-Field Transitions (SFT)

Strong-field transitions take place at values of $x \approx 1$. There, at fixed magnetic field and frequency only transitions between single Zeeman states are possible. The transitions are π or σ transitions. Typical transitions are $1 \leftrightarrow 4$ for H, $2 \leftrightarrow 6$, and $3 \leftrightarrow 5$ for D. For these, being σ transitions, the field direction of B_1 is parallel to the static field B_0. The higher frequencies require single-loop or, more modern, RF-cavity designs.

As for the WFT semi-classical as well as quantum-mechanical design criteria for the adiabatic-fast passage method can be derived. The criteria—using again the ficitious spin-1/2 system to relate the real quantum system to the semi-classical conditions—are:

1. The adiabaticity condition: The condition as defined for the WFT case translates into a condition which for MFT/SFT depends on the transition itself and cannot be generalized. Therefore, only an example for the often-used SFT between states 3 and 5 of deuterium will be given:

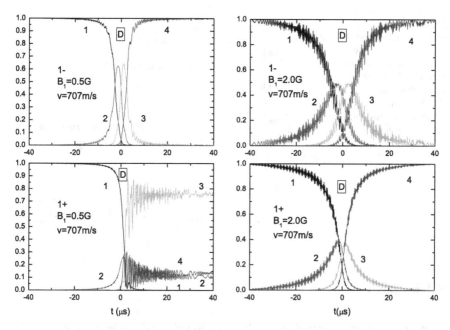

Fig. 8.17 Plot of the change of occupation numbers of Zeeman hyperfine states of deuterium starting with state 1 occupied only, through the WFT region. *Upper plots* are for negative, *lower* for positive field gradient. *Left*: $|B_1| = 0.05$ mT; *right*: $|B_1| = 0.2$ mT

$$B_1^2 \gg \frac{9}{2} \frac{\bar{v}}{\ell} \frac{\hbar}{\mu_B} \left[\left(x - \frac{1}{3} \right) \sqrt{1 - \frac{2}{3}x + x^2} \right] \Delta B \tag{8.73}$$

$\bar{v}$ is the average velocity of the atoms in the beam, ℓ the length of the RF interaction region.

2. The polarization condition: This translates into

$$B_1 \ll \frac{3}{\sqrt{2}} \left(x - \frac{1}{3} \right) \Delta B \tag{8.74}$$

3. The resonance condition is trivial.

These conditions are different for each of the possible transitions. Figure 8.18 shows an example for the design criteria for a deuterium transition from state 3 to state 5 at $B_0 = 10$ mT, $\Delta B_0 = 0.7$ mT, a transition frequency $\nu = 350$ MHz, average velocities of the atoms corresponding to temperatures of 100 and 300 K, and a length of the RF region of 3 cm. Quantum-mechanical calculations for the strong-field transitions $2 \leftrightarrow 6$, $3 \leftrightarrow 5$ for deuterium, and $2 \leftrightarrow 4$ for hydrogen have been used to study the dependence of the transition probabilities on different parameters in Hasuyama et al. [35, 36]. In Beijers [32] numerical calculations for WFT and SFT in hydrogen are compared to analytical expressions from the generalized Landau-Zener-Stückelberg theory [31, 37–39].

Fig. 8.18 Plot of the
semi-classical criteria for the
choice of the SFT working
point for the deuterium σ
transition $3 \leftrightarrow 5$ at
$B_0 = 10\,\text{mT}$ and
$\nu = 350\,\text{MHz}$. The
numbering corresponds to
the list of criteria above

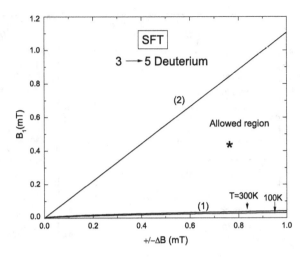

8.4 Ionizers

In order to convert polarized atomic beams into polarized ion beams a number of
different schemes have been developed:

- Electron-Bombardment ionizers
- ECR ionizers
- CBS ionizers

8.4.1 Ionizers: Electron-Bombardment and CBS Designs

8.4.1.1 Electron Bombardment Ionizers

The cross-section for ionization of hydrogen by electron impact has a maximum
near $E_e = 70\,\text{eV}$ (see Fite and Brackmann [40], Kieffer and Dunn [41]). Figure 8.19
shows the results of several authors. The classical strong-field ionizer makes use of
a long ionization path by three measures:

- Spiraling of the electrons, emitted by a cathode wire and accelerated by a positive
  grid or ring electrode, along a strong magnetic field of >0.1 T.
- Long ionization volume.
- Multiple use of electrons by reflection from a repulsive electric field at the end of
  the volume, serving at the same time as extraction field for the ions.

The electron space-charge depression has to be compensated by injecting the
electrons at voltages much higher than 70 eV. The prototype of this ionizer was
developed by Glavish [42] and has been used in many positive-ion sources. When

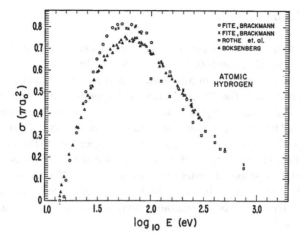

Fig. 8.19 Cross-section of ionization by electron impact of atomic hydrogen with a maximum near 70 eV. (From Kieffer and Dunn [41]; ©; [1966] by APS, New York)

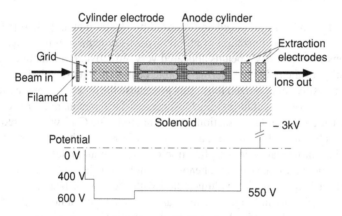

Fig. 8.20 Scheme of a Glavish-type electron-bombardment ionizer

negative ions were required—such as for tandem Van-de-Graaff accelerators—an additional charge exchange in alkali vapor had to follow. A recent application was in LS polarimeters for use on atomic beams [43, 44]. Figure 8.20 shows the scheme of such a device which has an ionization efficiency of up to $5 \cdot 10^{-3}$ and in which later the background current was reducecd by adding getter pumping around the ionization volume [45].

8.4.1.2 ECR Ionizers

The high ionization efficiency of electron-cyclotron-resonance ionizers was exploited in some polarized-ion sources [46, 47]. The ECR principle is to ionize the polarized beam by electrons accelerated in a plasma created by an intense RF discharge in a strong magnetic field. The RF frequency corresponds to the electron-cyclotron resonance and is therefore coupled to the magnetic field. At the magnetic field of $>100\,$mT optimal RF frequencies around 3.8 GHz are necessary with an RF power up to several 100 W. The fields are shaped to confine the electrons to the ionizing region and to extract the ions efficiently and without depolarization. The plasma discharge has to be maintained stably at rather low pressures $<1 \cdot 10^{-6}$ mbar which has been achieved by bleeding inert gas like N_2 into the discharge volume. One advantage of ECR ionizers is their small beam emittance. Ionization efficiencies of up to $6 \cdot 10^{-3}$ have been reported.

8.4.1.3 CBS Ionizers

This type of source was proposed and realized by Haeberli et al. [48, 49]. The very high cross-sections for ionization of atomic hydrogen/deuterium into negative ions in collisions with neutral Cs beams appeared very attractive. Even more attractive is the ionization of the polarized thermal atomic beams by intense colliding beams of negative or positive unpolarized ions. This is because the cross-sections are larger by about two orders of magnitude at very low energies due to resonant charge exchange. In Fig. 8.21 the relevant cross-sections are compared. The CBS with Cs requires an energetic Cs beam of about 45 keV, as shown in Fig. 8.22 with the charge-exchange cross-section into negative ions as function of Cs-beam and relative energies. It is, however, mandatory that such a Cs beam with high intensity could be produced and guided into a long ionization volume filled with the atomic beam. The Cs^+ ions with currents of many mA (up to 15 mA) are extracted from a hot tungsten surface ionizer button with about 45 keV energy, then neutralized efficiently in Cs vapor. Due to the high current density nearly complete space-charge neutralization takes place thus allowing beams of very high brilliance to reach the charge-exchange cell and produce good-geometry neutral beams which could be transported over distances of meters, see e.g. Fig. 8.23. Such CBS sources were successfully built at Madison, Brookhaven, Seattle, and, finally, for COSY-Jülich. Figure 8.24 shows he original design of a collaboration of three university groups (Erlangen, Bonn, and Cologne). A number of essential features made this source a superior device with high output, high polarization, reliability and long-time running capability. These were: 20 ms pulsing of the atomic beam by 20 ms feeding of gas input as well as that of N_2 and O_2, synchronous pulsing of the Cs beam [51], cutting down Cs consumption, electron-bombardment heating of the tungsten surface-ionizer button, a modified negative-ion extraction scheme etc.

The use of resonance ionization by low-energy, but high-intensity beams of H^-, D^-, H^+, or D^+ meets the difficulty of high space charge which so far restricts

Fig. 8.21 Cross-sections for ionization of H or D into positive or negative ions by charged H (or D), neutral Cs, and electron beams

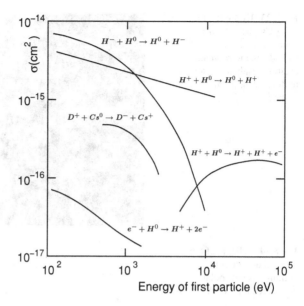

Fig. 8.22 Cross-section for ionization of H or D into negative ions by a neutral Cs beam. From [50]

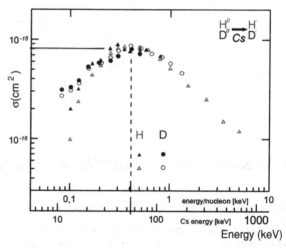

the realization only as pulsed systems with very short pulses (μs). Plasma ionizers with very high ionization efficiencies have been developed. Though very high peak pulse currents (up to 50 mA) have been reached the average number of polarized particles per unit time remains relatively small. The CBS with Cs is in principle a DC source, but in connection with pulsed accelerators such as COSY/Jülich with long (20 ms) pulses the performance is much improved by pulsing the source.

Fig. 8.23 Example of a
1 mA Cs$^+$ beam emitted
from the hot tungsten
cathode on the *right* and
exciting residual-gas atoms
to fluorescence while exiting
to the *left*

Fig. 8.24 Scheme of the
original design of the
colliding-beams polarized
ion source for COSY-Jülich

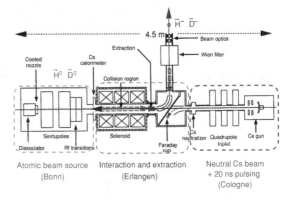

8.4.2 Sources for Polarized ^{6,7}Li and ^{23}Na Beams

In the past atomic-beam polarized ion sources for $\overrightarrow{^{6,7}Li}$ ($I = 1$ and $3/2$) and
$\overrightarrow{^{23}Na}$ ($I = 3/2$) beams have been developed (see e.g. Ebinghaus et al. [52], Bartosz
et al. [53], for a survey see e.g. several contributions in Kondo et al. [54]). In the
Spin-3/2 cases the complete description of the polarization requires tensor moments
t_{kq} up to rank $k = 3$. The construction of these tensor moments from spin operators
is described in Darden [55], see also Sect. 3.6.9.

The techniques of producing the atomic beams are different from the hydrogen
case: atoms are evaporated from an oven and ionization can be done by surface
ionization on heated W metal. In the first such sources Stern–Gerlach separation
magnets have been used for spin-state separation. Later optically pumped sources
were developed (e.g. OPPLIS at Florida State University, see [56–58]).

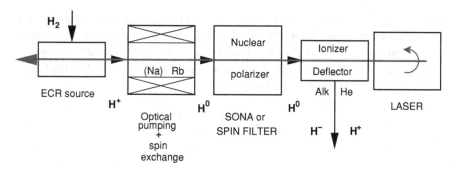

Fig. 8.25 Schematic showing the principles of optically pumped polarized-ion sources

8.4.3 Optically Pumped Polarized Ion Sources (OPPIS)

The principle used here is the same as that of polarized targets applying the optical pumping of alkali vapors (especially rubidium) and transfer of the high electronic polarization by collisions to the ground-state atoms and nuclei of H or D (spin-exchange method SEOP). The relevant wavelength (795 nm) is in the near infrared and pumping can be done by different lasers (e.g. Ti:Sapphire), but high-power laser diodes (diode arrays) have recently become available and are most convenient. Figure 8.25 shows a schematic of the principal functions of an OPPIS. Different functions such as neutralization of the injected H^+ beam, optical pumping of Rb, and spin-exchange collisions may occur in an integral vessel in a common magnetic field region. A source of this type, using the charge-exchange reaction

$$H^0 + He \rightarrow H^+ + He + e \qquad (8.75)$$

for positive ion output, developed originally at TRIUMF/Vanvouver, is being used at the RHIC accelerator at BNL/Brookhaven. It has superior properties: ionization efficiency up to 0.8, DC $\overrightarrow{H^-}$ currents of up to 15 mA (pulsed about 25 mA) at high polarizations [59].

8.5 Physics of the Lambshift Source LSS

8.5.1 The Lambshift

Lambshift = energy difference between the $2S_{\frac{1}{2}}$ and the $2P_{\frac{1}{2}}$ states [60], explained only by quantum electrodynamics. For hydrogen this shift (without a magnetic field) is about 1057 MHz or $4.38 \cdot 10^{-6}$ eV. The lifetime for the transition $2S_{\frac{1}{2}} - 2P_{\frac{1}{2}}$, due to the very small energy difference, is about 20 years. A dipole transition (E1) to the $1S_{1/2}$ ground state is forbidden ($I = 0 \rightarrow I = 0$), as is the corresponding quadrupole transition (E2) ($J = 0 \rightarrow J = 0$). A magnetic dipole transition (M1) is allowed and its lifetime was calculated to be about 2 days [61]. The main contribution comes from a two-quantum transition with $\tau = \frac{1}{7}s$. An electric field reduces the lifetime of the $2S_{\frac{1}{2}}$ and increases that of the $2P_{\frac{1}{2}}$ state via the Stark effect, which mixes states of different parity (i.e. here the parity is not a good quantum number). Following Lamb & Retherford the lifetime of the $2S$ state is

$$\tau_S \cong \tau_P \frac{\hbar^2(\omega^2 + \frac{\gamma^2}{4})}{|V|^2} \qquad (8.76)$$

with

$$\tau_P = \text{lifetime of the P state} = 1.595 \cdot 10^{-9} \, \text{s} \qquad (8.77)$$

$$\hbar\omega = \Delta E$$
$$= \text{energy separation between S and P state (field dependent).} \qquad (8.78)$$

$$\gamma = 1/\tau_P \qquad (8.79)$$

$$V = \langle \varphi_S | eEr | \varphi_P \rangle$$
$$= \text{matrix element of the dipole transition} \qquad (8.80)$$

which, for small electric fields (≤ 100 V/cm), is approximately

$$\tau_S = \tau_P (475/E)^2. \qquad (8.81)$$

8.5.2 Level Crossings and Quench Effect

In the picture of the fine structure (FS) the Stark effect mixes states with $\Delta m_J = 1$, $\Delta \pi = +$, i.e. (in the historical nomenclature of Lamb and Retherford) the states α and f, β and e, respectively. Because the states β and e cross at a magnetic field of about 57.5 mT, the transition probability there becomes maximal. The lifetime of β becomes shorter with smaller ΔE (the transition probability (perturbation calculation!) contains $(\Delta E)^2$ in the denominator). $\tau_S(\alpha)$ increases with B because of increasing state separation, while $\tau_S(\beta)$ has a minimum near 57.5 mT. The lifetime of the S state is empirically given by the formula:

$$\tau_S = \frac{1.13}{E^2} \left[(574 \pm B)^2 + 716 \right] \text{ns} \qquad (8.82)$$

For $E = 15$ V/cm one obtains e.g. $\tau_S(\alpha)/\tau_S(\beta) = 1850$. For a hydrogen beam with 500 eV ($3.1 \cdot 10^7$ cm/s) practically all atoms have decayed into the state β after 6.5 cm, but only 3.5% of the α states (see Figs. 8.2 and 8.26). In this way an atomic beam is obtained which is about 96% polarized in the electronic spin. The HFS Zeeman splitting leads to four (or nine, resp.) crossings around 57.5 mT of which

Fig. 8.26 Lifetimes of the n = 2 Zeeman states as functions of the magnetic field for two electric quenching field strengths

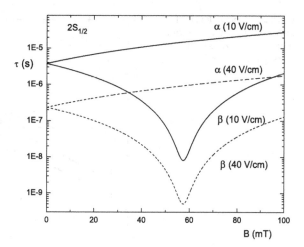

two (for H) or three (for D) can undergo Stark-effect quenching (see also Fig. 8.34). If the beam is (adiabatically) transported and ionized in a weak magnetic field a nuclear polarization of half of the theoretical value of the electronic polarization (and a correspondingly polarized proton or deuteron beam) results.

8.5.3 Enhancement of Polarization

There are two ways to enhance and change the nuclear polarization in the metastable beam. One is the use of a non-adiabatic (fast) transition with a change of the occupation of the Zeeman states, the SONA transition scheme [62]. The other is the use of a spin-filter [63] in which a combination of a longitudinal magnetic field, a transverse static electric field and an RF field lead to the transmission of single HFS states. These methods result in polarization values close to the theoretical maxima.

Figure 8.27 depicts the Breit-Rabi diagram for D with a sudden field reversal via a zero-crossing. Depending on the degree of adiabaticity of the crossing, the occupation of the Zeeman states follows different "trajectories" on the Zeeman levels (see Sect. 8.3.4). The practical realization of the LSS will be addressed below.

8.5.4 Examples of the Polarization Calculation for Different Modes of the LSS

Figure 8.28 shows different modes of operation of the LSS with one (for vector polarization of protons or deuterons) or two quenching processes (for deuteron tensor polarization).

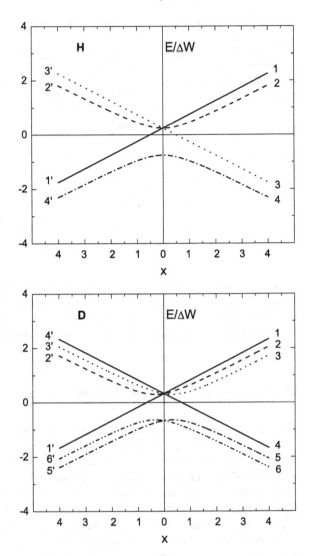

Fig. 8.27 Zeeman levels of hydrogen and deuterium with (non-adiabatic zero-field crossing/field inversion

8.5.5 Hydrogen

Sona magnet I is set at 57.5 mT, Sona magnet II at 25.0 mT, thus only one quenching process can occur. Basically only the hyperfine components 2 and 3 are occupied.

$$p^* = \frac{N_+ - N_-}{N_+ + N_-} = -\frac{1}{2}\left(1 + \frac{x}{\sqrt{1+x^2}}\right) \tag{8.83}$$

Depending on the ionizer field strength B this results in:

Fig. 8.28 Different SONA modes of operation of the LSS for H (*top*) and D (*bottom*)

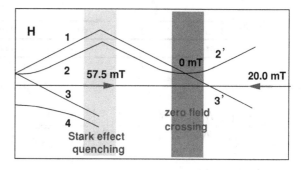

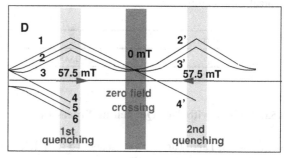

B (mT)	0	6	18	∞
p^*	−0.5	−0.842	−0.971	−1

8.5.6 Deuterium

8.5.6.1 Mode with One Quenching Process

Sona magnet I is set at 57.5 mT, Sona magnet II at about 20 mT. The Sona transition leads to the occupation basically only of components 2, 3, and 4.

- Vector polarization

$$p^* = \frac{N_{+1} - N_{-1}}{N_{+1} + N_0 + N_{-1}} = -\frac{1}{3} \cdot \frac{1}{2}\left[\frac{x + \frac{1}{3}}{\sqrt{x^2 + \frac{2}{3}x + 1}} + \frac{x - \frac{1}{3}}{\sqrt{x^2 - \frac{2}{3}x + 1}} + 2\right]$$

(8.84)

- Tensor polarization

    Depending on the magnetic field strength in the ionizer region one obtains e.g:

B (mT)	0	6.0	18.0	∞
p^*	$-\frac{1}{3}$	0.658	-0.666	$-\frac{2}{3}$

$$p^*_{zz} = \frac{N_{+1} + N_{-1} - 2N_0}{N_{+1} + N_0 + N_{-1}} = -\frac{1}{3} \cdot \frac{3}{2} \left[\frac{x + \frac{1}{3}}{\sqrt{x^2 + \frac{2}{3}x + 1}} - \frac{x - \frac{1}{3}}{\sqrt{x^2 - \frac{2}{3}x + 1}} \right] \quad (8.85)$$

Typical values are:

B (mT)	0	6	18	∞
p^*_{zz}	$-\frac{1}{3}$	-0.004	-0.0002	0

8.5.6.2 Mode with Two Quenching Processes

Sona magnets I and II are each set at about 57.5 mT, leading to an occupation basically only of components 2 and 3. With

$$N_0 = 1/2 \left(2 + \frac{x + \frac{1}{3}}{\sqrt{x^2 + \frac{2}{3}x + 1}} - \frac{x - \frac{1}{3}}{\sqrt{x^2 - \frac{2}{3} + 1}} \right) \quad (8.86)$$

and

$$N_{+1} = 1/2 \left(1 - \frac{x + \frac{1}{3}}{\sqrt{x^2 + \frac{2}{3}x + 1}} \right) \quad (8.87)$$

and

$$N_{-1} = 1/2 \left(1 + \frac{x - \frac{1}{3}}{\sqrt{x^2 - \frac{2}{3}x + 1}} \right) \quad (8.88)$$

follows:

B_{Ion} mT	p^*	p^*_{zz}
0	0	-1
.5	-0.144	-0.933
1.5	-0.348	-0.679
4.0	-0.470	-0.520
6.0	-0.486	-0.506
∞	$-\frac{1}{2}$	$-\frac{1}{2}$

Fig. 8.29 Proton polarization as function of the ionizer field after one quench

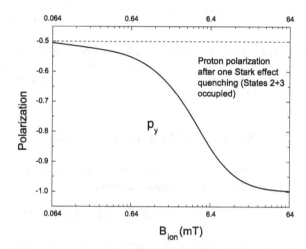

Fig. 8.30 Deuteron vector and tensor polarization as functions of the ionizer field with one (1) or two (2) quench processes

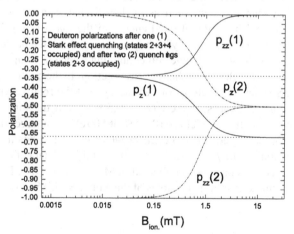

Figures 8.29 and 8.30 show the dependence of the proton or deuteron polarizations on the magnetic field at the ionizer location.

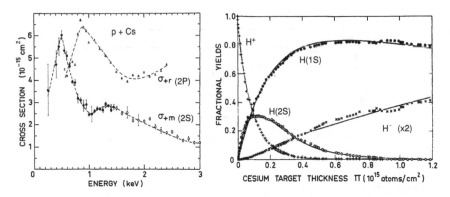

Fig. 8.31 Charge-exchange cross-sections of protons in Cs vapor into the metastable 2S state and the 2P state as functions of the energy (*left*) and relative contributions in the beam after passage through Cs vapor as functions of the areal thickness of Cs (*right*). The figures are from Pradel et al. [66]; ©; [1974] by APS, New York

8.5.6.3 Production of the Beam of Metastables

H and D atoms in the 2 S state may be produced by electron impact (see e.g. Lamb [60]), but the charge-exchange reaction

$$H^+ + Cs^0 \rightarrow H^0(1S, 2S) + Cs^+ + Q \tag{8.89}$$

is much more efficient. The primary positive ion beam is produced in a standard ion-source system such as an RF discharge, a duoplasmatron or even ECR ion source. The required fixed energy of the ions of 500 eV for H^+ or 1 keV for D^+ and therefore relatively high space-charge limit the beam current that can be injected into a charge-exchange cell containing Cs vapor of appropriate density. Figure 8.31 shows the charge-exchange cross-sections to the neutral ground and to the metastable 2S states as functions of the energy, and the relative yields as functions of the Cs target thickness. For Cs $Q = 0.50$ eV, and the ionization energy of 3.89 eV is very small. The yield is 10–15% at 500 eV for a target thickness of $5 \cdot 10^{-3}$ Torr cm. The measured fraction of metastables in the full beam of neutral particles (1S, 2S) amounted to $f_{max} = 0.430 \pm 0.03$. Additional data of these reactions from a number of authors have been collected in Morgan et al. [64]. Figure 8.32 shows the scheme of the source of metastables of the Köln LSS LASCO (see also Bechtold et al. [65]).

8.5.7 Production and Maximization of the Beam Polarization

In order to obtain maximum values of the polarization with a LSS, in analogy to the ABS transitions between hyperfine states are induced. However, because the

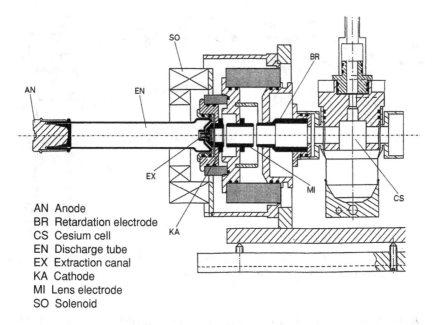

AN Anode
BR Retardation electrode
CS Cesium cell
EN Discharge tube
EX Extraction canal
KA Cathode
MI Lens electrode
SO Solenoid

Fig. 8.32 Detail of the Köln LSS LASCO

(metastable) beam, in comparison to the ground-state atomic beam, is "fast" none of the usual adiabatic RF transitions can be used, but either non-adiabatic transitions (SONA transitions) or a SPINFILTER. This leads to two possible schemes for the construction of the LSS, depicted in Fig. 8.33.

8.5.7.1 SONA Transition

Basic idea: non-adiabatic transition in a "rapidly" sign-changing magnetic field (zero crossing) [62]. "Rapidly" means: The field change happens in time intervals short against $1/\nu_L = 2\pi/\omega_L$, from which the condition

$$1/B(dB/dt) \gg \omega_L/2\pi = (\gamma/2\pi)B, \tag{8.90}$$

i.e.

$$dB/dt \gg (\gamma/2\pi)B^2 \tag{8.91}$$

is derived. In this case the atoms stay in their respective Zeeman HFS states while the field is reversed, see also Fig. 8.16 and the discussion there. Thus the original state 1 becomes state $1' \equiv 4$, leading to a theoretical nuclear polarization of 100% instead of 50%. There is a critical volume: The non-adiabaticity condition is always fulfilled

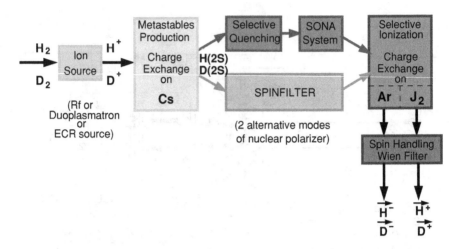

Fig. 8.33 Realization of the LSS with two principles: SONA method (*upper*) and SPINFILTER (*lower*). Negative as well as positive ions can be obtained

for $B=0$, i.e. on the beam axis as long as the field has a gradient at all. Away from the beam axis the field can only be $\neq 0$. Therefore, there is a critical beam radius beyond which this condition is not fulfilled. In addition, any superimposed external fields B_t (e.g. the earth's magnetic field) require a minimum gradient of the longitudinal field B_0. This condition is numerically [62]:

$$4B^2{}_t/14G \cdot cm < B'_0 < 14G \cdot cm/r^2 \tag{8.92}$$

Its derivation uses the fact that a transverse field B_t corresponds to a lateral shift of the location of zero-crossing according to

$$\Delta r = 2B_t/B'_0 \tag{8.93}$$

Example: With $B_t = 0.5\,G$, $B'_0 = 2\,G/cm$ the condition reads:

$$1/14 < 0.2\,mT/cm < 14/r^2 \tag{8.94}$$

which for $r = 1\,cm$ is well satisfied. The field gradient must not be so large that the transition from a weak-field situation $B < B_{crit}$ to the situation of a strong field $B > B_{crit}$ becomes non-adiabatic. This can be ensured by making the SONA region large enough. Technically this has been realized by two properly shaped magnetic fields of opposite polarity in z direction at such a distance that the gradient condition along the beam path is fulfilled. In addition, external fields must be screened or compensated for.

8.5.7.2 Spin Filter

The theory of the spin filter [63, 67] is somewhat complicated because its function rests on the simultaneous interaction of three states:

- The $1S_0$ ground state,
- the metastable $2S_{1/2}$ state, and
- the short-lived $2P_{1/2}$ state.

The Breit-Rabi diagram Fig. 8.34 illustrates (for H) the simultaneous interactions. Near the level crossings the β states are quenched, i.e. decay rapidly into the 1S ground state. The RF transition depopulates the substate $\alpha 2$ while the substate $\alpha 1$ is constantly repopulated from one β substate. After exiting the spin filter only one hyperfine substate remains populated. The choice of the magnetic field value at fixed RF frequency (or vice versa) determines which state is being transmitted. For deuterium the interactions are analogous. The interactions are realized by the static longitudinal magnetic field of a solenoid of about 57.5 mT, which must be quite homogeneous, a static electric quenching field realized by segmenting the RF cavity into quadrants and applying a DC voltage to an opposing quadrant pair, and an electric RF field with a frequency of $\nu = 1.60975$ GHz in the TM_{010} mode in a resonator cavity. The spin-filter setup is illustrated in Fig. 8.35. More details can be found in Engels [43], Engels et al. [44], Trützschler [68], Franke [69], Weske [70]. The spin-filter principle has advantages over the SONA principle, at least for deuterium. They derive from the fact that single hyperfine components can be selected and transmitted whereas the usual SONA scheme transmits two states. Therefore, only with the spin filter the theoretical maximum values of the polarization between $p_{ZZ} = +1$ and -2 can be obtained together with the possibility to change the sign of the polarization. On the other hand the intensity is reduced, as compared to the SONA scheme. Therefore, the figure of merit $p^2 \cdot I$ has to be evaluated for each scheme, and in general, the use of a spin filter may not be useful for protons, also in view of the simpler operation of the SONA scheme, whereas for the deuteron tensor polarization a doubling of the figure of merit was proven experimentally.

The function of a spin filter is illustrated by Fig. 8.36, which shows the polarization and the transmitted intensity (current) of the deuterons as functions of the spin-filter magnetic field keeping the E field and the RF frequency constant.

8.5.7.3 Selective Ionization of the (Polarized) Beam of Metastables

This is achieved by a quasi-resonant charge-exchange process

$$H(2S) + Ar \rightarrow H^- + Ar^+ \tag{8.95}$$

A similar charge exchange leads to positive polarized ions [71, 72]:

$$H(2S) + I_2 \rightarrow H^+ + I_2^-. \tag{8.96}$$

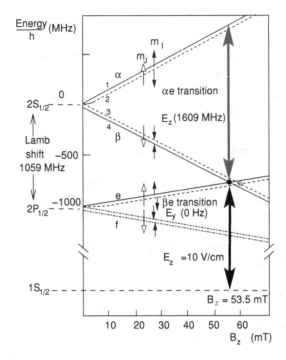

Fig. 8.34 2S/2P hydrogen hyperfine-Zeeman states diagram with transitions caused by the static and the RF electric fields

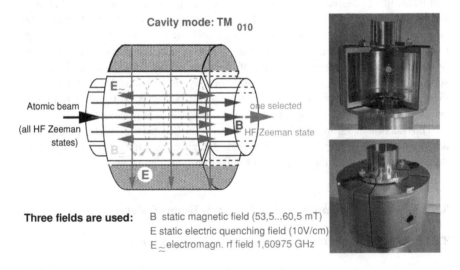

Fig. 8.35 Scheme of a spin filter with the relevant fields and photographs

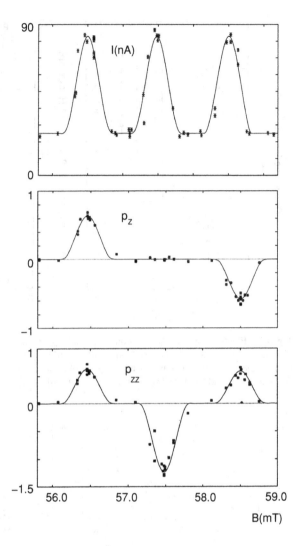

Fig. 8.36 Transmitted current and polarization of the deuteron hyperfine components appearing at three different values of the spin-filter magnetic field B [69]

Fig. 8.37 shows the high value and weakly resonant behavior of the cross section σ_{2S-} for negative-ion formation from metastables as compared to σ_{1S-} from ground-state atoms (left). The right part of the figure shows the strongly energy-dependent (relative) H$^-$ ionization yields of these processes, especially the high selectivity of the metastable relative to the ground state. For the 2S state an ionization energy of at least $10.19 + 0.75 = 10.94$ eV (i.e. the excitation energy of the 2S state plus the binding energy B.E. of the electron in H$^-$) is required. Argon has an ionization energy of 15.8 eV and is therefore especially suited. The selectivity 2S/1S is almost 100%, the ionization yield is near 1%. The high (and resonant) selectivity of this process is accentuated by the measurement of Donnally and Sawyer [73] where the yield of negative ions after metastable production by a proton (similarly for deuterons at

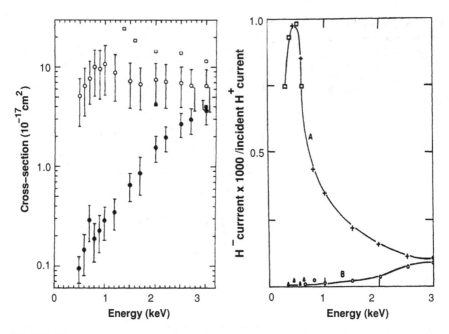

Fig. 8.37 Electron capture cross-sections σ_{1S-} and σ_{2S-} on argon for ground-state and metastable H atoms as functions of the energy, from Roussel [76] (*left*) and $H^+ \rightarrow H^-$ charge-exchange (i.e. through the entire system as used in a LSS) yield of ground-state and metastable H atoms on argon, from Donnally and Sawyer [73] (*right*); ©; [1977,1968] by APS, New York

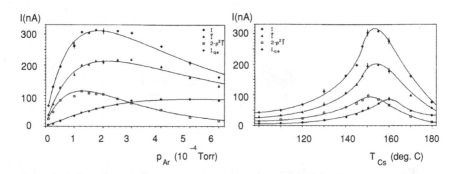

Fig. 8.38 Example for the optimization of the argon-cell pressure (*left*) and Cs cell temperature (*right*) with respect to $p^2 \cdot I$. For comparison ion currents are also shown [74]

the same velocity) beam in Cs vapor (curve A) is compared to that after Stark-effect quenching of the metastables (curve B). In practice, all source parameters must be optimized with respect to the maximum figure of merit $p^2 \cdot I$. Figure 8.38 gives an example in which two of many parameters (e.g. quenching-field voltage, positive-beam velocity, magnetic Sona field strength, etc.) have been optimized: the Ar cell gas pressure and Cs cell temperature [74].

The final limitations of the LSS are at least twofold. They consist in the necessity of working at given low energies for the production of metastables which limits the current of H^+/D^+ ions that can be injected into the Cs charge-exchange region due to space charge. Attempts to neutralize this space charge have been made by injecting electrons from a heated filament into the positive-ion beam with <1.5 μA polarized beam intensity [75]. Another limiting factor is the relativistic quenching by the electric field $\vec{E} = \gamma\vec{v} \times \vec{B}$ generated by the fast motion of the metastables in magnetic fields such as the SONA fields.

8.6 Spin Rotation in Beamlines and Precession in a Wien Filter

Each polarization facility needs the capability of preparing the polarization at the source and/or in the beamline such that the absolute value of the polarization as well as its orientation in space can be optimized, the latter freely chosen. The direction of the polarization vector coincides with the direction of the principal axis of the polarization tensor (for deuterons).

8.6.1 Spin Rotation in Beamlines

Normally the beamlines contain deflection magnets and electric deflection fields. The latter do not influence the direction of the polarization in a space-fixed coordinate system, but of course the angle between the polarization vector and the direction of motion of the beam may change. In magnetic fields (such as from analyzing and switching magnets) the spin polarization precesses except when $\vec{p}$ is parallel to $\vec{B}$.

- The precession of a nuclear spin $\vec{I}$ in a magnetic field can be described by the classical relation between the torque and angular momentum $\vec{M} = \vec{\mu} \times \vec{B} = g_I \mu_N \frac{1}{\hbar}(\vec{I} \times \vec{B})$ and $\vec{M} = \frac{d\vec{I}}{dt}$:

$$\frac{d\vec{I}}{dt} = \frac{q}{2m}g_I \cdot (\vec{I} \times \vec{B}) = g_I \cdot \mu_N \frac{1}{\hbar} \cdot (\vec{I} \times \vec{B}). \tag{8.97}$$

For the proton (m_p = proton mass)

$$g_I = \frac{m}{m_p} \cdot \frac{e}{q} \cdot g_{Landé}. \tag{8.98}$$

$\frac{d\vec{I}}{dt}$ is oriented perpendicular to $\vec{I}$, i. e. only the direction but not the absolute value of $\vec{I}$ changes. Likewise $\frac{d\vec{I}}{dt}$ is perpendicular to $\vec{B}$ and the spin precesses around $\vec{B}$. The Larmor precession period T is given by

$$\frac{2\pi}{T} = \left|\frac{d\vec{I}}{dt}\right| \frac{1}{|\vec{I}|\sin\phi} = g_I \cdot \mu_N \frac{1}{\hbar} |\vec{I}||\vec{B}| \sin\phi \frac{1}{|\vec{I}|\sin\phi} \qquad (8.99)$$

The precession is independent of the angle between the spin and the magnetic field and occurs with the circular *Larmor frequency*

$$\omega_L = \frac{g_I \mu_N B}{\hbar} \qquad (8.100)$$

with the nuclear magneton

$$\mu_N = \frac{e\hbar}{2m_p} = 5.05 \cdot 10^{-27} \text{ J/T.} \qquad (8.101)$$

- The Landé g factors for the proton and deuteron are: $g_p = 5.586$ and $g_d = 0.857$.
- The magnetic moments of the proton, the deuteron, and triton are positive. Therefore, for a positive beam (such as on the high-energy side of a tandem Van-de-Graaff accelerator) the sense of spin rotation in these cases is the same as that of the magnetic deflection. For the polarized beams from negative-ion sources the opposite is true, which also has to be taken into account in a Wien filter.
- The spin Larmor precession and the deflection in magnetic fields, which is described by a *cyclotron motion* with $\omega_C = (q/m)B$, are proportional to each other and coupled together. The change of the polarization direction relative to the beam is the difference between the angles of precession and of deflection.
- Although nuclear reactions are preferably described in a coordinate system where the y axis is along $\vec{k}_{in} \times \vec{k}_{out}$ and is therefore different for each detector, the preparation of the polarization in the entrance channel may better be done in a beam-fixed coordinate system with a y axis vertical in space, a z axis attached to the beam direction (which may change in beam-deflection devices) and the x axis forming a right-handed system with both. An azimuthal angle ϕ of the polarization vector is counted starting from the x axis.
- In a deflection magnet (with B field in y direction) the precession occurs in the x–z plane and the change of the spin polar angle $\Delta\beta$ (measured from the z axis) with fixed azimuthal angle ϕ is (for positive beams) is:

$$\Delta\beta = \Delta\theta_L - \Delta\theta_C = \left(g\frac{m}{2m_p} - 1 \right) \Delta\theta_C, \qquad (8.102)$$

i.e. for protons:

$$\Delta\beta = 1.793 \cdot \Delta\theta_C \qquad (8.103)$$

and for deuterons:

$$\Delta\beta = -0.143 \cdot \Delta\theta_C. \qquad (8.104)$$

For negative beams:

$$\Delta\beta = -\Delta\theta_L - \Delta\theta_C = \left(-g\frac{m}{2m_p} - 1\right)\Delta\theta_C, \qquad (8.105)$$

8.6.2 Spin Rotation in a Wien Filter

In order to set the polarization direction to any desired angle at the target a spin rotation device is required, preferably already at the source where beam velocities are still low. For this purpose a Wien (velocity) filter which is rotatable around the beam axis on the source is especially suited. With the ion beam of velocity $\vec{v}$ in z direction, an electric field $\vec{E}$ in x, and a magnetic field $\vec{B}$ in y direction the filter transmits ions fulfilling the condition

$$v = \frac{E}{B} \qquad (8.106)$$

for an ideal reference beam, i.e. one on the central z axis. For an extended beam with finite emittance (i.e. with particles having transverse momentum components) the above condition cannot be fulfilled for all particles simultaneously. This results in some (small) spreading of final spin directions in the beam, i.e. depolarization. This effect can be reduced by having a beam cross-over in the center of the device.

In Ohlsen [77], Buballa [78] the changes of the spin orientation by deflecting fields and precession in a Wien filter are described. Two questions arise with respect to spin precession:

- Rotations are generally described by three Euler angles. However, the direction of a spin vector is completely determined by two parameters, i.e. in polar coordinates a polar angle β and an azimuthal angle ϕ. Thus, one Euler angle is redundant (i.e. can only enter as a phase factor) and the other two are uniquely connected with β and ϕ.
- It can be shown that the description of the quantum-mechanical rotation of spinors via action of rotation matrices on spin operators (such as represented by Pauli matrices) is entirely equivalent to a classical 3×3 rotation matrix acting on spin vectors [78].

If we describe the orientation of a spin unit vector in the beam-fixed coordinate system x, y, and z defined above, by polar coordinates

$$\hat{S} = \begin{pmatrix} \sin\beta\cos\phi \\ \sin\beta\sin\phi \\ \cos\beta \end{pmatrix} \qquad (8.107)$$

then after a general rotation by polar and azimuthal angles α and ψ we have

$$
\begin{pmatrix} \hat{S}_{x'} \\ \hat{S}_{y'} \\ \hat{S}_{z'} \end{pmatrix} = \begin{pmatrix} \sin\beta'\cos\phi' \\ \sin\beta'\sin\phi' \\ \cos\beta' \end{pmatrix} = \begin{pmatrix} \cos\alpha\cos\psi & \cos\alpha\sin\psi & \sin\alpha \\ -\sin\psi & \cos\psi & 0 \\ \sin\alpha\cos\psi & \sin\alpha\sin\psi & \cos\alpha \end{pmatrix} \cdot \begin{pmatrix} \sin\beta\cos\phi \\ \sin\beta\sin\phi \\ \cos\beta \end{pmatrix}
$$

$$(8.108)$$

For an accelerator system where all rotations occur around the y axis, i.e. where $\psi = 0$ the rotation matrix simplifies to

$$
\begin{pmatrix} \cos\alpha & 0 & \sin\alpha \\ 0 & 1 & 0 \\ -\sin\alpha & 0 & \sin\alpha \end{pmatrix}.
$$

$$(8.109)$$

By applying the inverse of this matrix

$$
\begin{pmatrix} \cos\alpha & 0 & -\sin\alpha \\ 0 & 1 & 0 \\ \sin\alpha & 0 & \cos\alpha \end{pmatrix}.
$$

$$(8.110)$$

to an arbitrary desired spin orientation at the target the necessary setting of the Wien filter magnetic field and azimuthal orientation can be calculated. Under the action of a vertical field B of length L along z the spin (polarization) vector of the particles precesses in the x–z plane by a polar angle proportional to $|B|$ and inversely proportional to v.

$$
\beta_{prec} = \frac{g_I \mu_K}{\hbar} \frac{(B \cdot L)_{eff}}{v}
$$

$$(8.111)$$

In the general case the azimuthal orientation of the field $\vec{B}$ determines the azimuthal angle of the polarization around the z axis. The Wien filter $\vec{E}$ field is adjusted for maximum transmission of the polarized beam. At the same time the device acts as a velocity filter (and mass filter for ions of the same energy). After also taking into account subsequent changes of the spin orientation by other deflecting fields (analyzer or switching magnets etc.) where the total spin change is described by multiplication of the rotation matrices of all devices, every desired spin orientation at the target position may be achieved. Such systems have e.g. been built and used at Basel [79], Los Alamos [77], Köln [74, 78], COSY/Jülich [80], and others. It is evident that beam deflection by electric fields—although it does not affect the spins—also changes the orientation of the polarization vector when described with respect to the z axis (direction of motion).

Figure 8.39 shows precession curves of the vector polarization of protons and deuterons as well as of the tensor polarization of the deuterons. The polarizations were measured with specially developed polarimeters using the reactions ^{3}He(d,p)^{4}He, ^{4}He(p,p)^{4}He, and ^{4}He(d,d)^{4}He after the acceleration by a tandem Van-de-Graaff accelerator and display the variation of the polar angle β of the polarization. By rotating the Wien filter around the beam axis an arbitrary variation of the azimuthal angle ϕ may be achieved. Thus, after taking into account other ion-optical elements of the beam-transport system such as deflection magnets etc., any direction

Fig. 8.39 Precession curves of the polarization (values were normalized to 100%) as functions of the current of the Wien-filter magnetic field (the corresponding electric field is then fixed for a given particle velocity and is found by maximum beam transmission). (**a**) Vector polarization of protons, (**b**) Tensor polarization of deuterons, (**c**) Vector polarization of deuterons [74]

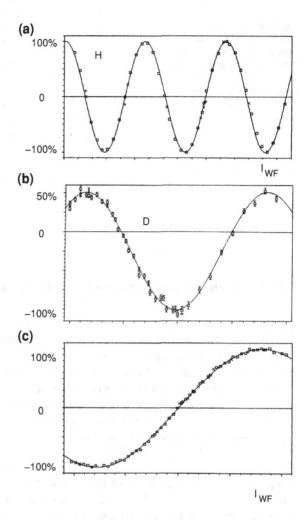

of the polarization vector (or tensor) at arbitrary locations of experimental setups can be selected. The stigmatic optical properties of Wien filters are those of cylindrical lenses—they focus a beam with circular cross-section into a line—can be improved by adding an element with similar properties as an electric (or magnetic) quadrupole singlet [70]. With a typical tensor polarimeter using four or five detectors in a 3D arrangement not only the amount of polarization, but also the direction of the polarization vector in space can be determined (see e.g. Engels [81], Czerwinski [82]). Thus also the function of the spin-precession device used for spin orientation can be checked independent of the absolute value of the polarization. The Wien filter has to be calibrated with respect to the magnetic-field coil current vs. the precession angle, exactness of the azimuthal orientation and possible zero-offset, caused e.g. by remanent magnetism, environmental fields, and non-linearities due to saturation effects in the iron.

Fig. 8.40 Schematic of a storage cell with typical pressure distribution along the cell with polarized atoms entering from the *top*. The projectile beam crosses the cell horizontally

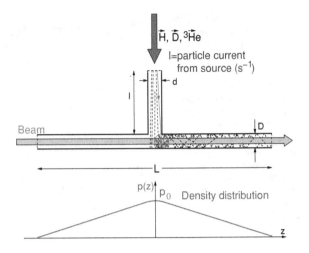

8.7 Polarized (Gas) Targets and Storage Cells

This lecture note will be restricted to gas (polarized-beam) targets. The techniques for producing polarization in solids are quite different from those required here and are partly based on solid-state and low-temperature physics exceeding the scope of this text, and the reader is referred to the special literature.

The physics and techniques of polarized (jet or beam) targets are basically the same as those of the ABS. No ionization is needed, and the basic requirement is maximum (areal) density of polarized atoms in the interaction region. The temperature of the beam is important because of the relation $\rho = j/v \propto jT^{-1/2}$.

Due to the low density of such jets (on the order of $\approx 1 \cdot 10^{12}$ cm^{-2}) when compared to solid targets the real potential of polarized jets as targets lies in applications at storage rings/synchrotrons such as COSY where the accelerated beams traverse the target many (typically 10^6) times. The additional use of storage cells increases the target density. A storage cell consists of a tube into which the polarized beam is fed from the side, the atoms undergo a number of collisions with the walls of the tube before being pumped away after exiting the ends of the tube. The important feature when using polarized beams is the degree of depolarization during the wall collisions. It turned out that hundreds of collisions are possible especially when the walls of the storage cell have been covered with some special material with low recombination/depolarization rates. In the molecular-flow regime (i.e. when the mean free path is large compared to the vessel dimensions) the outflow is mainly governed by the dimensions of the tube and the temperature of the beam atoms. The ratio of this outflow and the incoming beam intensity lead to a pressure equilibrium with a triangular pressure distribution along the tube and an effective density increase of up to two orders of magnitude. Two comprehensive references cover the status of polarized gas targets up to 2005, see Steffens and Haeberli [83] and Rathmann [84]. The areal density of such a storage cell can be calculated from the conductances of

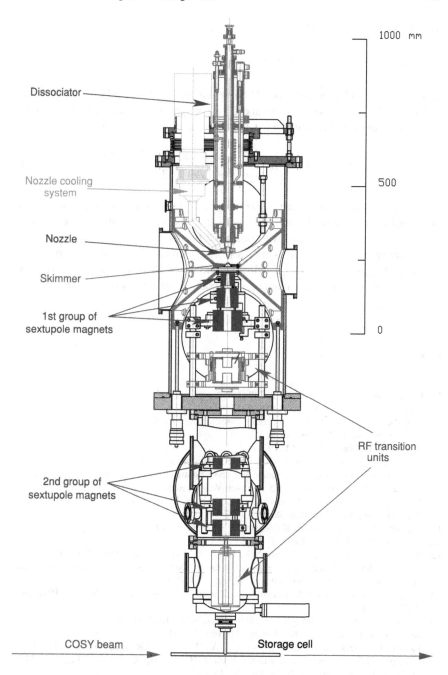

Fig. 8.41 ANKE polarized-atom target at COSY-Jülich (Engels R, private communication, 2010)

the tube segments: Fig. 8.40

$$n = \int \rho_{H^0} dL \approx \frac{N_{H^0} L}{2C_{tot}} \tag{8.112}$$

with N_{H^0} the number of polarized H atoms entering the side tube, $L = L_1 + L_2$ the total length of the storage tube, and L_3 the length of the feeding tube. The total conductance of the entire cell is the sum of the conductances of the three segments:

$$C_{tot} = 3.81 \cdot 10^3 \sqrt{\frac{T}{m}} \sum_i \frac{d_i^3}{L_i + 1.33 d_i} \left[\frac{cm^3}{s} \right] \tag{8.113}$$

with T the absolute temperature of the tube, m the mass of the beam atoms in *amu*, the d_i and L_i the diameters and lengths of the tube segments in cm. Nearly $n \approx 1 \cdot 10^{14} cm^{-2}$ has been reached with polarized beams. Figure 8.41 shows a modern ABS feeding polarized particles into a storage cell at COSY-Jülich. A proof-of-principle of applying the storage-cell principle to a *CBS polarized-ion source* was realized and showed that there is appreciable potential of development for this method [23, 24].

References

1. Gerlach, W., Stern, O.: Z. Physik **9**, 349 (1922)
2. Frazer, R.G.J.: Proc. Roy. Soc. A **114**, 212 (1927)
3. Uhlenbeck, G.E., Goudsmit, S.A.: Naturwissenschaften **13**, 953 (1925)
4. Gerlach, W., Stern, O.: Z. Physik **9**, 353 (1922)
5. Thomas, L.H.: Nature **117**, 514 (1926)
6. Friedrich, B., Herschbach, D.: Phys. Today **53**, December (2003)
7. Feynman, R.P., Leighton, R.B., Sands, M.: The Feynman Lectures on Physics, vol. II chapter 35 and vol. III chapters 5, 6, 11, 35. Addison-Wesley, Reading (1965)
8. Mitter, H.: Quantentheorie. BI-Hochschultaschenbuch, Mannheim (1976)
9. Breit, G., Rabi, I.I.: Phys. Rev. **38**, 2082 (1931)
10. Clausnitzer, G., Fleischmann, R., Schopper, H.: Z. Physik **144**, 336 (1956)
11. Clausnitzer, G.: Z. Physik **153**, 600 (1959)
12. Rudin, H., Striebel, H.R., Baumgartner, E., Brown, L., Huber, P.: Helv. Phys. Acta **34**, 58 (1961)
13. Huber, P., Meyer, K.P. (eds.): Proceedings International Symposium on Polarization Phenomena of Nucleons, Basel 1960, Helv. Phys. Acta Suppl. VI. Birkhäuser, Basel (1961)
14. Singy, D., Schmelzbach, P.A., Grüebler, W., Zhang, W.Z.: Nucl. Instrum. Methods Phys. Res. A **278**, 349 (1989)
15. Berney, J.C., Dick, L., Kubischta, W.: Helv. Phys. Acta **59**, 578 (1986)
16. Koch, N., Steffens, E.: Rev. Sci. Instrum. **70**, 1631 (1999)
17. Paetz gen. Schieck, H., Busch, C.E., Keane, J.A., Donoghue, T.R.: In: [85] p. 810 (1971)
18. Donoghue, T.R., McEver, W.S., Paetz gen. Schieck, H., Volkers, J.C., Busch, C.E., Doyle, M.A. Sr., Dries, L.J., Regner, J.L.: In: [87] p. 840 (1976)
19. Halbach, K.: Nucl. Instrum. Methods **169**, 1 (1980)
20. Vasiliev, A., Paetz gen Schieck, H. et al.: Rev. Sci. Instrum. **71**, 3331 (2000)
21. Kumada, M., Fujisawa, I., Hirao, Y., Iwashita, Y., Endo, M., Aoki, M., Kohda, T., Bolshakova, I., Holyaka, R.: IEEE Trans. Appl. Supercond. **12**(1), 129 (2002)

22. Tenckhoff, G., Emmerich, R., Paetz gen. Schieck, H.: Verhdl. DPG, HK18.13 (2003)
23. Emmerich R.: Ph. D. thesis, Universität zu Köln (2001) http://kups.ub.uni-koeln.de/volltexte/2007/2081
24. Emmerich, R., Paetz gen. Schieck, H.: Nucl. Instrum. Methods Phys. Res. A **586**, 387 (2008)
25. Abragam, A., Winter, J.: Phys. Rev. Lett. **1**, 375 (1958)
26. Iwashita, Y., Tajima, Y., Ichikawa, M., Nakamura, S., Ino, T., Muto, S., Shimizu, H.M.: Nucl. Instrum. Methods Phys. Res. A **586**, 73 (2008)
27. Ramsey, N.: Molecular Beams. Oxford University Press, New York (1956)
28. Majorana, E.: Nuovo Cimento **9**, 43 (1932)
29. Paetz gen. Schieck, H.: Nucl. Instrum. Methods Phys. Res. A **587**, 213 (2008)
30. Haeberli, W.: Ann. Rev. Nucl. Sci. **17**, 373 (1967)
31. Philpott, R.J.: Nucl. Instrum. Methods Phys. Res. A **259**, 317 (1987)
32. Beijers, J.P.M.: Nucl. Instrum. Methods Phys. Res. A **536**, 282 (2005)
33. Glavish, H.: In: [85], p. 267 (1971)
34. Oh, S.: Nucl. Instrum. Methods **82**, 189 (1970)
35. Hasuyama, H., Kanda, Y., Katase, A., Wakuta, Y.: Nucl. Instrum. Methods **207**, 475 (1983)
36. Hasuyama, H., Wakuta, Y.: Nucl. Instrum. Methods Phys. Res. A **260**, 1 (1987)
37. Landau, L.D.: Phys. Z. Sowjetunion **2**, 46 (1932)
38. Zener, C.: Proc. Roy. Soc. A **137**, 696 (1932)
39. Stückelberg, E.C.G.: Helv. Phys. Acta **5**, 369 (1932)
40. Fite, W.L., Brackmann, R.T.: Phys. Rev. **112**, 1141 (1958)
41. Kieffer, L.J., Dunn, G.H.: Rev. Mod. Phys. **38**, 1 (1966)
42. Glavish, H.: Nucl. Instrum. Methods **65**, 1 (1968)
43. Engels, R.: Ph.D. thesis, Universität zu Köln. http://kups.ub.uni-koeln.de/volltexte/2003/777 (2002)
44. Engels, R., Emmerich, R., Ley, J., Tenckhoff, G., Paetz gen. Schieck, H.: Rev. Sci. Instrum. **74**, 4607 (2003)
45. Engels, R., Emmerich, R., Grigoriev, K., Paetz gen. Schieck, H., Mikirtytchiants, M., Rathmann, F., Sarkadi, J., Seyfarth, H., Tenckhoff, G., Vasilyev, A.: Rev. Sci. Instrum. **76**, 053305 (2005)
46. Clegg, T.B., Hooke, W.M., Crosson, E.R., Lovette, A.W., Middleton, H.L., Pfutzner, H.G., Sweeton, K.A.: Nucl. Instrum. Methods Phys. Res. A **357**, 212 (1995)
47. Friedrich, L., Huttel, E., Schmelzbach, P.A.: Nucl. Instrum. Methods Phys. Res. A **272**, 906 (1988)
48. Haeberli, W.: Nucl. Instrum. Methods **62**, 355 (1968)
49. Haeberli, W., Barker, M.D., Gossett, C.A., Mavis, D.G., Quin, P.A., Sowinski, J., Wise, T.: Nucl. Instrum. Methods **196**, 319 (1982)
50. Annual Report, p. 51, Strahlenzentrum Universität Giessen (1981)
51. Eggert, M., Paetz gen. Schieck, H.: Nucl. Instrum. Methods Phys. Res. A **453**, 514 (2000)
52. Ebinghaus, H., Holm, U., Klapdor, H.V., Neuert, H.: Z. Phys. **199**, 68 (1967)
53. Bartosz, E. et al.: Phys. Lett. B **488**, 138 (2000)
54. Kondo, M., Kobayashi, S., Tanifuji, M., Yamazaki, T., Kubo, K.-I., Onishi, N. (eds): Proceedings of 6th International Symposium on Polarization Phenomena in Nuclear Physics, Osaka 1985, Suppl. J. Phys. Soc. Jpn. 55 (1986)
55. Darden, S.E.: In: [85] p. 39 (1971)
56. Mendez, A., Myers, E.G., Kemper, K.W., Kerr, P.L., Reber, E.L., Schmidt, B.G.: Nucl. Instrum. Methods Phys. Res. A **329**, 37 (1993)
57. Cathers, P.D., Green, P.V., Bartosz, E.E., Kemper, K.W., Marechal, F., Myers, E.G., Schmidt, B.G.: Nucl. Instrum. Methods Phys. Res. A **457**, 509 (2001)
58. Weintraub, W.D., Cathers, P.D., Green, P.V., Bartosz, E.E., Kemper, K.W., Marechal, F., Myers, E.G., Schmidt, B.G.: Nucl. Instrum. Methods Phys. Res. A **491**, 349 (2002)
59. Zelenski, A.: In: [86] (2011)
60. Lamb, W.E. Jr., Retherford, R.C.: Phys. Rev. **79**, 549 (1950)

61. Breit, G., Teller, E.: Astrophys. J. **91**, 215 (1940)
62. Sona, P.G.: Energia Nucleare **14**, 295 (1967)
63. McKibben, J.L., Lawrence, G.P., Ohlsen, G.G.: Phys. Rev. Lett. **20**, 1180 (1968)
64. Morgan, T.J., Olson, R.E., Schlachter, A.S., Gallagher, J.W.: J. Phys. Chem. Ref. Data **14**, 971 (1985)
65. Bechtold, V., Friedrich, L., Ziegler, P., Aniol, R., Latzel, G., Paetz gen. Schieck, H.: Nucl. Instrum. Methods **150**, 407 (1978)
66. Pradel, P., Roussel, F., Schlachter, A.S., Spiess, G., Valance, A.: Phys. Rev. A **10**, 797 (1974)
67. Ohlsen, G.G., McKibben, J.L., Los Alamos Scientific Lab. Report LA-3725 (1967)
68. Trützschler, A.: Diploma thesis, Universität zu Köln (1994, unpublished)
69. Franke, C.: Diploma thesis, Universität zu Köln (1995, unpublished)
70. Weske C.: Diploma thesis, Universität zu Köln (2003, unpublished)
71. Knutson, L.D.: Phys. Rev. A **2**, 1878 (1970)
72. Brückmann H., Finken D., Friedrich L.: Nucl. Instrum. Methods **87**, 155 (1970) and: In: [85] p. 823 (1971)
73. Donnally, B.L., Sawyer, W.: Phys. Rev. Lett. **15**, 439 (1965)
74. Reckenfelderbäumer, R.: Diploma thesis, Universität zu Köln (1989, unpublished)
75. Sagara, K.: Few-Body Systems **48**, 59 (2010)
76. Roussel, F.: Phys. Rev. A **16**, 1854 (1977)
77. Ohlsen, G.G.:Los Alamos Scientific Lab. Report LA-4451 (1970)
78. Buballa, M.: Diploma thesis, Universität zu Köln (1999, unpublished)
79. Paetz gen. Schieck, H., Huber, P., Petitjean, Cl., Rudin, H., Striebel, H.R.: Helv. Phys. Acta **40**, 414 (1967)
80. Weidmann, R., Glombik, A., Meyer, H., Kretschmer, W., Altmeier, M., Eversheim, P.D., Felden, O., Gebel, R., Glende, M., Eggert, M., Lemaître, S., Reckenfelderbäumer, R., Paetz gen. Schieck, H.: Rev. Sci. Instrum. **67**, 1357 (1996)
81. Engels, R.: Diploma thesis, Universität zu Köln (1997, unpublished)
82. Czerwinski, A.: Diploma thesis, Universität zu Köln (1999, unpublished)
83. Steffens, E., Haeberli, W.: Rep. Progr. Phys. **66**, 1887 (2003)
84. Rathmann, F.: Nucl. Instrum. Methods Phys. Res. A **536**, 235 (2005)
85. Barschall, H.H., Haeberli, W. (eds.): Proceedings of 3rd International Symposium on Polarization Phenomena in Nuclear Reactions, Madison 1970, University of Wisconsin Press (1971)
86. Rathmann, F., Ströher, H. (eds.): Proceedings of International Spin Conference (SPIN2010), Jülich 2010. Published as Open Access by IOP Conference series (2011)
87. Grüebler, W., König, V. (eds.): Proceedings of 4th International Symposium on Polarization Phenomena in Nuclear Reactions, Zürich 1975, Birkhäuser, Basel (1976)

Chapter 9
Polarization by Optical Pumping

9.1 Principles

Optical pumping has been applied mainly to polarized alkali and H beams, see Sect. 8.4, using the method of spin exchange. For polarized ^{3}He targets metastability as well as spin exchange are used.

9.2 Polarization of ^{3}He

The polarization of ^{3}He has been achieved very successfully by optical pumping methods. Polarized ^{3}He gas targets have been used in many nuclear- and particle-physics experiments. An interesting feature is that polarized ^{3}He is to a great extent equivalent to a nearly fully (above 80%) polarized neutron. However, corrections due to the neutron being bound in a three-nucleon environment, from microscopic theory, are needed. Thus, $\overrightarrow{^3\text{He}}$ targets have been used not only in low-energy nuclear experiments studying e.g. the ^{3}He $(d,p)^4$He [1] and ^{3}He $(p,p)^3$He [2] reactions, but also in particle-physics experiments such as DESY-HERMES [3] in order to learn about the spin structure of the neutron in electron scattering. Another interesting application is that a beam of neutrons, after passing through a polarized $\overrightarrow{^3\text{He}}$ target becomes a beam of highly polarized neutrons, see Sect. 10.3. Due to the strong dependence of the total cross section on the (relative) spin directions this system acts as an efficient spin filter. The development of the field in basic nuclear science and in interesting applications can be followed in proceedings of special conferences and workshops, see [4–6].

Two basic schemes have developed which shall be described here, the spin-exchange ("SEOP") and the metastability-exchange ("MEOP") method. The spin structure of the He atom with zero electronic spin is different from that of the hydrogen istopes and Stern–Gerlach or Lambshift schemes cannot be employed easily because only

H. Paetz gen. Schieck, *Nuclear Physics with Polarized Particles*,
Lecture Notes in Physics 842, DOI: 10.1007/978-3-642-24226-7_9,
© Springer-Verlag Berlin Heidelberg 2012

the very small nuclear Zeeman splitting in a magnetic field could be used. One advantage, on the other hand, is that the relaxation times of the nuclear polarization are long due to the weakness of the interaction of the nuclei with the electronic environment (i.e. in collisions with other atoms, walls etc.). Thus optical pumping of gaseous ^{3}He in glass vessels is the preferred method. Both methods have their specific advantages and disadvantages.

9.2.1 Polarization by Metastability Exchange

Here the optical pumping is done starting from the excited 2S level of the ^{3}He atom into the hyperfine substates of the the 2P level. Figure 9.1 shows the relevant parts of the atomic level scheme of ^{3}He together with that of ^{4}He. Thus, the atoms have to be excited into the 2S state by a weak radiofrequency discharge ($\approx$40 MHz) across the ^{3}He sample. It turns out that the efficiency (measured by the figure of merit $p^2 \cdot p_{gas}$) of the process is maximal at a ^{3}He gas pressure p_{gas} of about 0.4 kPa. Early efforts to compress the polarized ^{3}He gas without substantial loss of polarization have been undertaken using liquid Ga [7] and using a piston pump with an Al_2O_3 ceramic piston in a Ti vessel or a Toepler compression pump with mercury [8], a system which has been perfected at Mainz [9] and used for medium- and high-energy nuclear and particle-physics experiments [10] as well as in the application in medical physics, see below in Chap. 13.

The wavelength of this transition is 1083 nm and is in the near infrared. Before the advent of lasers in this range gas-discharge lamps filled with ^{4}He emitting this wavelength corresponding to a strong 2P $\rightarrow$ 2S transition had to be used, resulting in polarization values around 20%. Today lasers of the krypton arc lamp-pumped Nd:LNA type (see e.g. [11]) and broadband ytterbium-doped tunable fiber lasers (see e.g. [12]) allow high spectral intensities and corresponding high values of the nuclear polarization of up to 70% or higher.

The pumping light has to be right- or left-circularly polarized by a conventional linear polarizer plus $\lambda/4$ plate combination. The high nuclear polarization in the atomic metastable 2S state is transferred to the ground state by collisions with unpolarized ground-state atoms, a process called metastability exchange.

The relaxation times of the polarization are very long but also very sensitive to impurities especially to those from magnetic materials. Therefore, thoroughly cleaned glass vessels of special materials have to be used for the pumping cells and possible target cells where the same is true for beam entrance and exit windows. Since for accelerator experiments and medical applications the low density is very restrictive, the compression schemes discussed have provided target pressures above 1 bar at high polarization values 60% and long relaxation times. Target cells have been separated from the pumping cells. For medical applications (see below) large samples are routinely produced and can be transported for long times in portable magnetic fields.

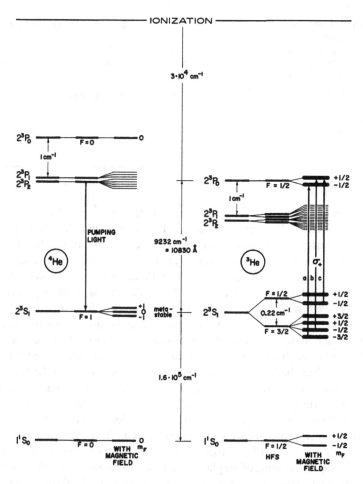

Fig. 9.1 Atomic level schemes of ^{3}He and ^{4}He. The figure shows how—before the advent of lasers—the light emitted by a transition in ^{4}He was used to pump transitions in ^{3}He leading to a net nuclear polarization of the ^{3}He nuclei. The ^{4}He light was emitted from a gas-discharge excited by RF. For more details see Refs. [1, 13, 14]

9.2.2 Spin Exchange

Here the optical pumping is done in alkali vapors such as rubidium in the ground state (no RF excitation necessary). Thus the sample can be pumped at much higher density (pressures of several bar, i.e. 300–800 kPa) than in the metastability-exchange case. Diode laser arrays with output power of up to 100 W, delivering linearly polarized light, and $\lambda/4$ plates provide the necessary intense circularly polarized radiation. Figure 9.2 depicts a simplified scheme of the optical-pumping process using circularly polarized laser light thus depleting one spin substate and filling up the other.

Fig. 9.2 Optical-pumping process using right-circularly polarized light from a diode laser with $\lambda = 795$ nm enriching the $m_S = +1/2$ state and depleting the $m_S = -1/2$ state leading to a positive electronic polarization of the Rb atoms

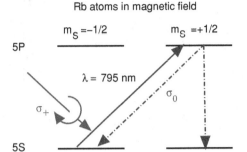

The Rb vapor has to be maintained at an optimum density, i.e. at $\approx 185°$C by heating. Pumping time is typically a day at 800 hPa of ^{3}He with 1% admixture of N_2 [15]. The excited Rb atoms emit unpolarized photons which can excite surrounding Rb atoms thus causing depolarization ("radiation trapping"). By adding molecular nitrogen which absorbs energy into rotational and vibrational excitations this effect can be reduced substantially to only about 5% of the excited atoms emitting photons. The Rb vapor is mixed with ^{3}He gas and N_2 with densities $n(Rb) \ll n(N_2) \ll n(^3He)$. The final polarization is given by the laser power and the spin-destruction rate from collisions. The polarization of the Rb atoms is transferred to the ^{3}He atoms by Rb–^{3}He collisions through a hyperfine interaction. The transfer rate is, however, quite small (≈ 3%). Thus the transfer process is quite inefficient and—in the presence of depolarizing collisions—requires long pumping times. These depolarization effects are from ^{3}He–^{3}He collisions (dipole interaction) and ^{3}He-wall collision, sensitive to magnetic impurities, and from magnetic field gradients. The addition of K as an additional vapor for which the spin-exchange rates with Rb and ^{3}He are much higher than for Rb–^{3}He alone ("SEOP" process) makes the pumping process more efficient ("hybrid cells"). The coating of the glass vessels of low-paramagnetic aluminosilicate glass (GE180) with suitable inorganic compounds such as $Al(NO_3)_3 \cdot 9H_2O$ improved the final ^{3}He polarization. Values higher than 60% have been reached. Pumping times are typically several tens of hours, much longer than with MEOP. For use as practical targets under beam irradiation conditions the functions of pumping and target cells are separated. The required two cells, connected by a tube, or two separate cells—one for pumping, the other as target cell—must be in a very homogeneous magnetic field (Helmholtz coils) and maintained at elevated temperatures. Special glass composition (aluminosilicate, borosilicate) for the pumping and target vessels are necessary, and the thin beam entrance and exit windows required for low-energy experiments cause problems. Making them from low-relaxation glass (as was done for MEOP) but stable against the higher pressure used with SEOP results in unacceptable thicknesses. On the other hand, no thin foils with small relaxation rates comparable to those of the special glasses were found. Thus, the choice of SEOP versus MEOP has to weigh better polarization combined with lower target density n and an additional compression device against faster relaxation, longer pumping times, but higher target density. Recent developments of polarized ^{3}He targets applied to low-energy nuclear physics experiments with their special problems as well as high-pressure developments are described in [2, 15–17].

9.3 Ion Sources for Polarized ^{3}He Beams

Up to 1984 three operational $\overrightarrow{^3\text{He}}$ sources connected to an accelerator were built and used for nuclear-physics experiments (at Birmingham, Laval, and Rice/Texas A&M). For example, the Birmingham source was of the Lambshift type, but with spin-state selection in the hydrogen-like ^{3}He$^+$ (2S) state produced from ^{3}He^{++}. Thus the beam intensities and/or polarizations achieved were quite low, see e.g. Ref. [18]. In view of the recent significant progress of optical-pumping techniques a number of proposals for sources of polarized ^{3}He beams have been made, but were not realized so far.

References

1. Leemann, Ch., Bürgisser, H., Huber, P., Rohrer, U., Paetz gen. Schieck, H., Seiler, F.: Helv. Phys. Acta **44**, 141 (1971)
2. Daniels, T.V., Arnold, C.W., Cesaratto, J.M., Clegg, T.B., Couture, A.H., Karwowski, H.J., Katabuchi, T.: Phys. Rev. C **82**, 034992 (2010)
3. DeSchepper, D. et al.: Nucl. Instrum. Methods Phys. Res. A **419**, 16 (1998)
4. Tanaka, M. (ed.): Proceedings of the 7th International RCNP Workshop on Polarized ^{3}He Beams and Gas Targets and Their Applications "HELION97", Kobe 1997. North Holland, Amsterdam (1997)
5. Heil, W. (ed.): Proceedings of the International Conference on Polarized ^{3}He ("HELION02"), Oppenheim 2002, on CD-ROM (2003)
6. JCNS Workshop on Trends in Production and Applications of Polarized ^{3}He, Ismaning 2010, IOP J. of Physics: Conference Series (2011)
7. Szaloky, G., Huber, P., Leemann, Ch., Rohrer, U., Seiler, F.: Helv. Phys. Acta **43**, 745 (1970)
8. Timsit, R.S., Daniels, J.M.: Can. J. Phys. **49**, 545 (1971)
9. Otten, E.: Europhysics News Jan./Febr. p. 16 (2004). http://www.europhysicsnews.org
10. Heil W.: http://www.physik.uni-mainz.de/helion02 and: In: [5]
11. Gentile, T.R., McKeown, R.D.: Phys. Rev. A **47**, 456 (1993)
12. Tastevin, G., Grot, S., Courtade, E., Bordais, S., Nacher, P.J.: Appl. Phys. B: Lasers and Optics **78**, 145 (2004)
13. Leemann, Ch., Bürgisser, H., Huber, P., Rohrer, U., Paetz gen. Schieck, H., Seiler, F.: Ann. Phys. (N.Y.) **66**, 810 (1971)
14. Huber, P., Leemann, Ch., Rohrer, U., Seiler, F.: Helv. Phys. Acta **42**, 907 (1969)
15. Katabuchi, T., Buscemi, S., Cesaratto, J.M., Clegg, T.B., Daniels, T.V., Fassler, M., Neufeld, R.B., Kadlecek, S.: Rev. Sci. Instrum. **76**, 033503 (2005)
16. Kramer, K., Zong, X., Lu, R., Dutta, D., Gao, H., Quian, X., Ye, Q., Zhu, X., Averett, T., Fuchs, S.: Nucl. Instrum. Methods Phys. Res. A **582**, 318 (2007)
17. Karban, O., Oh, S., Powell, W.B.: Phys. Rev. Lett. **33**, 1438 (1974)
18. Ye, Q., Laskaris, G., Chen, W., Gao, H., Zheng, W., Averett, T., Cates, G.D., Tobias, W.A.: Eur. Phys. J. A **44**, 55 (2010)

Part IV
Methods

Chapter 10
Production of Polarization by Other Methods

10.1 Polarized Charged-Particle Beams from Nuclear Reactions

Before the advent of polarized-ion sources connected to accelerators with their marked advantages polarized particles had to be produced in nuclear reactions. This method has a number of disadvantages.

- Low intensity of the outging particle beams.
- Very bad "beam" quality due to energy and angular spread.
- Dependence on details of spin-dependent interactions.
- Dependence of the polarization on energies and angles of specific nuclear reactions.

10.2 Polarized Neutrons from Nuclear Reactions

For nuclear reactions induced by polarized neutrons in the low-MeV energy range polarized neutrons have to be produced in nuclear reactions ("double-scattering"). In order to obtain high neutron intensities thick targets are used, and high cross-sections as well as high polarization values are required, where the figure of merit $p_n^2 \cdot d\sigma(\Theta)/d\Omega$ has to be maximized (p_n is the neutron polarization). Figure 10.1 shows this for a number of relevant reactions for unpolarized incident projectiles and for polarization transfer from highly polarized incident particles from accelerated charged-particle beams out of a polarized-ion source. It is evident that the latter yield more than an order of magnitude higher f.o.m. For more details see Ref. [1]. Reference [2] gives an account of the latest improvements for the $D(\vec{d}, \vec{n})^3$He reaction where the polarization of the outgoing neutrons had to be measured by a suitable polarimeter, together with relevant references. The calibration of neutron polarimeters requires special provisions including the determination of the detection efficiency of neutron detectors.

H. Paetz gen. Schieck, *Nuclear Physics with Polarized Particles*,
Lecture Notes in Physics 842, DOI: 10.1007/978-3-642-24226-7_10,
© Springer-Verlag Berlin Heidelberg 2012

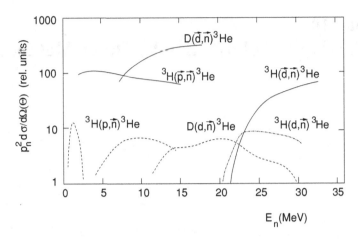

Fig. 10.1 Figure-of-merit of thick-target yields of three neutron-producing reactions. The *upper curves (solid lines)* are obtained using polarization transfer from highly polarized protons or deuterons at 0°, the *lower curves* are from reactions with neutron polarizations induced with unpolarized projectiles at angles $\Theta_{c.m.} = 45°$ *(dashed lines)* and $\Theta_{lab} = 30°$ *(dash-dotted line)*. The lines are interpolations adapted from data of several sources, the f.o.m. scale is relative to show the advantage of the polarization-transfer method

10.3 Spin Filtering: Interaction of Low-Energy Neutrons with Hyperpolarized ^{3}He

The basic idea of using polarized $\overrightarrow{^3\text{He}}$ as a spinfilter for neutrons was already discussed by Postma [3] at the Madison symposium [4]. The total (absorption) cross-section of very low energy neutrons, e.g., from a reactor has been shown to be very sensitive to the polarization of a ^{3}He target. The broad-band energy range (<1 keV) is that of cold, thermal, and epithermal neutrons. This effect lends itself to construct efficient polarizer as well as polarization analyzer systems, provided transmission type targets of high polarization and areal density can be obtained. Using the optical pumping methods (SEOP and MEOP plus compression) methods discussed above such systems have been built and applied for fundamental as well as applied research. It is basically the ^{3}He(n,p)^{3}H reaction which has a strong $J^\pi = 0^+$ state (resonance) at $E_n = -518$ keV, but with a width of ≈ 400 keV and can therefore be excited strongly by low-energy neutrons. The integrated cross-section of the ^{3}He(n,p)^{3}H reaction is $\sigma_r \approx 850\ b$ as compared to that of elastic scattering of $\sigma_{el} \approx 3\ b$ thus completely dominating the n + ^{3}He interaction. In addition, the spin structure of the reaction enforces that it can only take place between the neutrons with spins antiparallel to the $\overrightarrow{^3\text{He}}$ polarization causing strong absorption for these neutrons and high transmission for the others. This transmissions for the two groups are

$$\tau_\pm = \frac{1}{2}\ \exp[-(\sigma_0 \pm \sigma_p q_z)n\ell], \tag{10.1}$$

where $\sigma_0 = \frac{1}{2}(\sigma_+ + \sigma_-)$ and $\sigma_p = (\sigma_+ - \sigma_-)$ are the unpolarized and polarized cross-sections, q_z the polarization of ^{3}He, n the number density, and ℓ the length of the polarized ^{3}He sample. The neutron polarization is a function of the transmission ratio $\rho = \tau_+/\tau_-$

$$p_n = \frac{\rho - 1}{\rho + 1} \approx \tanh(\sigma_p q_z n \ell) \tag{10.2}$$

and the intensity of the transmitted beam is proportional to

$$\tau_n = \tau_+ + \tau_- \approx \tau_0 \cosh(\sigma_p q_z n \ell). \tag{10.3}$$

with $\tau_0 = \exp(-\sigma_0 n \ell)$ the transmission of an identical unpolarized sample. The ratio τ_0/τ_n is then a measure of the neutron polarization. Modern polarizer systems have reached neutron polarizations of practically 100%. The figure of merit which can be defined as $p_n^2 \tau_n$ depends of course on the ^{3}He polarization, but also on the "opaqueness" $\sigma \cdot n \cdot \ell$ and has maxima at different neutron energies for different density $(n \cdot \ell)$ values.

10.4 Spin Filtering for Polarized Antiprotons

The method of spin filtering by interactions between a polarized ensemble of particles and an unpolarized one is being discussed with respect to polarizing antiprotons. Experiments by the FILTEX collaboration at the TSR Heidelberg proved the buildup of a weak polarization ($p < 0.025$) of protons in collisions with polarized protons in a storage cell of high density. It appears that the mechanism of polarization buildup is a filtering process rather than a polarization transfer. At the FAIR facility being built at Darmstadt the use of (polarized) antiprotons is considered important in many respects, and the PAX collaboration has proposed a dedicated program of investigating spin filtering first at COSY-Jülich and then at the AD (antiproton decelerator) at CERN before designing a dedicated facility at FAIR. There, after producing polarized antiprotons by filtering in APR (antiproton polarizer ring) at up to 250 MeV the $\bar{p}$ will be stored and cooled at energies up to 3.5 GeV together with protons in CSR (cooler synchrotron ring) before being injected into HESR (high energy storage ring) with energies up to 15 GeV. Details can e.g., be found in Refs. [5–8].

References

1. Simmons, J.E., Broste, W.B., Donoghue, T.R., Haight, R.C., Martin, J.C.: Nucl. Instrum. Methods **106**, 477 (1971)
2. Roper, C.D., Clegg, T.B., Dunham, J.D., Mendez, A.J., Tornow, W., Walter, R.L.: Few-Body Systems **47**, 477 (2010)

3. Postma, H.: In: [4] p. 373 (1971)
4. Barschall, H.H., Haeberli, W. (eds.): Proceedings of the 3rd International Symposium on Polarization Phenomena in Nuclear Reactions. Madison 1970, University of Wisconsin Press, Madison (1971)
5. Rathmann, F.: In: [6] p. 106 (2008)
6. Barber D.P., Buttimore N., Chattopadhyai S., Court G., Steffens E. (eds.): Proceedings of the International Workshop on Polarized Antiprotons. Warrington, 2007. AIP Conf. Proc. 1008, New York (2008)
7. Crabb, D.G., et al. (eds.): Proceedings of the 18th International Symposium on Spin Physics, Charlottesville, 2008. AIP Conf. Proc. 1149, New York (2009)
8. Lenisa, P., Rathmann, F.: CERN Courier **50**(6), 21 (2010)

Chapter 11
Measurement of Polarization Observables

The primary quantity to be measured in nuclear reactions by any detector is always the intensity of reaction products. Contrary to polarization observables unpolarized cross-sections have no azimuthal-angle dependence. Therefore, the measurement of such observables must rely on this ϕ dependence which has been described in previous chapters. Wheras for an unpolarized cross-section only the two z directions of incoming and outgoing beams have physical meaning for polarization measurements the choice of coordinate systems is important for the proper description of the observables. The explicit ϕ dependence of a given system (reaction) relies on two facts:

- The spin structure of the system;
- The type of observable considered.

An example has been given for the spin-correlation cross-section of two spin-1 particles (see Fig. 5.2). Of course, to obtain any effect of the polarization in a nuclear reaction, the action of some spin-dependent force is required. The pure Coulomb force between two point charges (Rutherford scattering) is spin-independent, whereas an $\vec{\ell} \cdot \vec{s}$ force (such as required for the shell model of nuclear structure and also in the optical model of elastic nucleon scattering) leads to a left-right asymmetry of the outgoing particles when the incident particles are polarized or to a polarization of the outgoing particles with an unpolarized incident beam. (Both cases are related via the time-reversal invariance, as outlined above). This outgoing polarization can again be measured by another reaction showing a left-right asymmetry ("double-scattering") which is also the principle of polarization-transfer measurements when the incident beam is also polarized. Left-right asymmetries are the simplest case e.g. for spin-1/2 scattering which follows a $\cos \phi$ dependence. Tensor polarization of spin-1 particles produces e.g. $\cos 2\phi$ terms which can be measured either by appropriate arrangements of many detectors run simultaneously or by rotating one or few detectors around the z axis for different runs. It is clear that in this case the runs must be normalized to each other. Experimental (instrumental) asymmetries have to be avoided or cancelled by flipping the polarization vector and/or by interchanging detectors. These questions of systematic errors, spin flip etc. have been discussed

H. Paetz gen. Schieck, *Nuclear Physics with Polarized Particles*,
Lecture Notes in Physics 842, DOI: 10.1007/978-3-642-24226-7_11,
© Springer-Verlag Berlin Heidelberg 2012

in detail by Ohlsen et al. [1]. By considering the necessary measurement time for a given precision of an experiment a figure of merit was derived which is always the product of an intensity (or: cross-section; for polarized targets: density) times the square of a polarization component. All polarization experiments and devices should be optimized by maximizing this f.o.m. The quadratic dependence on the polarization makes clear that polarization values (or e.g. for polarimeter reactions: analyzing powers) should be as high as possible. The important question of absolute calibration of polarization and analyzing power is treated in Chap. 12.

Reference

1. Ohlsen, G.G. Keaton, P.W. Jr.: Nucl. Instrum. Methods **109**, 41 and 61 (1973)

Chapter 12
Polarimetry

Principally the polarization of beams is measured through a left-right or (for higher spins) through more complicated spatial asymmetries/anisotropies in a nuclear reaction. In any case the relevant analyzing powers have to be known with high accuracy since their errors enter the errors of the measured reaction directly. Therefore methods for absolutely calibrating analyzing powers are of particular importance here. In cases where no nuclear reaction can be used to measure the polarization, e.g. for measurements at a polarized source, in or after a storage cell, or on optically pumped polarized gas targets (such as ^{3}He, ^{129}Xe) properties of the electron shell have to be exploited. Via the hyperfine interaction the nuclear polarization can be deduced from the electron polarization.

In Lambshift polarized ion sources polarization can be measured using the "quench-ratio" method by comparing beam intensities after the spinfilter with and without quenching of the metastables by an electric field. For storage-cell targets a "Balmer" polarimeter was developed by which the H or D atoms excited by electron impact emitted Balmer fluorescence radiation, the circular polarization of which is a measure of the atomic polarization [1]. In optically pumped gas targets a nuclear-magnetic-resonance signal is induced where the signal strength is a measure of the polarization. Especially in the beginning of polarized-target physics the optical-absorption signal of the pumping radiation was measured [2–4] and allowed a reasonably good determination of the electron's polarization due to the fact that with increasing polarizaton the absorption tends to zero.

12.1 Absolute Methods

12.1.1 Time Reversal and Double Scattering

The classical method has been to perform a double-scattering experiment. The reciprocity theorem (time-reversal invariance) equates (among other observables) the analyzing power of the forward reaction (with polarized particles prepared in the

H. Paetz gen. Schieck, *Nuclear Physics with Polarized Particles*,
Lecture Notes in Physics 842, DOI: 10.1007/978-3-642-24226-7_12,
© Springer-Verlag Berlin Heidelberg 2012

entrance channel) with the polarization produced in the (time-)reversed reaction, initiated with unpolarized particles. For elastic scattering forward and reversed reactions are identical. Therefore, when producing an outgoing polarization in a scattering reaction with incident unpolarized particles and analyzing it by the same reaction the following relation (for spin-1/2 particles) holds for the second scattering:

$$\sigma_{pol} = \sigma_0(1 + p^2 \sin \Phi) = \sigma_0(1 + A^2 \sin \Phi) \tag{12.1}$$

The measurement of a left-right asymmetry then yields A or p, but not their sign. Both scatterings have to be measured at the same c.m. energies and the same scattering angles which will cause practical problems because of the energy losses, straggling etc. These problems can be overcome by additional related measurements at suitable energies. The sign of A (or p) can be obtained (for charged particles) from the interference term with the (calculable) Rutherford amplitude.

12.1.2 Analytical Behavior of the Scattering Amplitudes

Provided the scattering amplitudes change only continuously with energy Plattner et al. [5] have shown that ander certain conditions energies must exist at which the analyzing power assumes its maximum value. The following argument for this is used: One assumes that the real and imginary parts of the scattering amplitude $f(E)$ and $g(E)$, respectively are such that the (vector-) analyzing power is maximal and $= 1$. With the definition

$$A_y = 2\frac{\text{Im } (fg^*)}{|f|^2 + |g|^2} = 2\frac{\text{Re } (f^*(-ig))}{|f|^2 + |g|^2} \tag{12.2}$$

the conditions for maximum analyzing power are shown to be:

$$g = \pm if \rightarrow |f| = |g| = 1 \quad \text{and} \quad f \perp g, \tag{12.3}$$

i.e. f and g have a relative phase difference of $\pi/2$. Setting arbitrarily $f = e^{i\phi}$ to be real and $=1$ then: $g = \pm i$. Thus it is necessary to obtain experimental data in the vicinity of a large value of A_y in small angular and energy steps. Using a phase-shift (or scattering-amplitude) analysis a point must be searched fulfilling these conditions (provided such a point exists). A number of such points has been found e.g. for the "classical" analyzing reaction $^4\text{He}(\vec{p}, p)^4\text{He}$. The following table shows such points for this reaction:

$E_{p,lab}$	$\Theta_{c.m.}$
1.90 ± 0.02 MeV	$88.0 \pm 0.25°$
6.35 ± 0.04 MeV	$128.8 \pm 0.1°$
12.30 ± 0.04 MeV	$125.5 \pm 0.1°$

The authors suggest additional calibration points for the reaction $^4\text{He}(\vec{n}, n)^4\text{He}$ and $^3\text{He} - {}^4\text{He}$ elastic scattering ($A_y = -1$ for $^3\text{He}(\alpha, {}^3\text{He})^4\text{He}$ with $E_\alpha = 15.3\,\text{MeV}$, $\Theta_{^3\text{He},lab.} = 45°$).

For the vector and tensor analyzing powers of deuteron reactions a similar behavior was found [6] leading to absolute calibration points. *One example is*: The $^3\text{He}(\vec{d}, p)^4\text{He}$ reaction which has an absolute calibration point with $A_{yy} = 1$ at $E_d = 9\,\text{MeV}$ and $\Theta_{c.m.} = 27°$.

12.1.3 Calibration Points Due to a Special Spin Structure

Systems with a special spin structure may show analyzing powers which depend only on symmetries, not on the reaction dynamics. They assume certain values independent of energy and angle [7].

Examples are all reactions with the spin (and parity) structure

$$0^+ + 1^+ \rightarrow 0^+ + 0^+. \tag{12.4}$$

Here rotation and parity invariance restrict the analyzing powers to fixed values independent of energy and angle. As can be calculated e.g. with the program FATSON this spin structure allows only

$$A_y = A_{xz} = 0, \quad A_{xx} = A_{zz} = 1, \quad A_{yy} = -2A_{xx} - A_{yy} = 3 \tag{12.5}$$

or

$$T_{20} = \frac{1}{2}\sqrt{2}, \quad T_{11} = T_{21} = 0, \quad T_{22} = \frac{1}{2}\sqrt{3}, \tag{12.6}$$

respectively.

However, useful reactions with this spin structure are rare and often difficult to realize experimentally. A well-known example is the reaction:

$$^{16}\text{O}(\vec{d}, \alpha_1)^{14}\text{N}^*(2.31\,\text{MeV}) \tag{12.7}$$

to the first excited state of ^{14}N. This channel is isospin-forbidden and consequently has a very small cross-section. Nevertheless this reaction has been used for the definition of absolute calibration points and, with these, secondary polarimeter reactions have been calibrated. An example is the calibration of the $^4\text{He}(\vec{d}, d)^4\text{He}$ reaction [8]:

$$A_{yy} = -1.066 \pm 0.034 \quad \text{at} \quad E_d = 7.07\,\text{MeV}, \Theta_{lab} = 55° \tag{12.8}$$

In this way a number of additional secondary calibration points of the $^4\text{He}(\vec{d}, d)^4\text{He}$ reaction and the $^3\text{He}(\vec{d}, p)^4\text{He}$ reaction were obtained [9, 10], and others.

Also the reaction $^{12}C(\vec{d}, \alpha)^{10}B(2^+)$ at $0°$ with $A_{yy} = -1/2$, i.e. $A_{zz} = +1$ has been used for absolute polarimeter calibration [11].

Another example are reactions with the spin structure

$$\frac{1}{2} + 1 \rightarrow \frac{1}{2} + 0. \tag{12.9}$$

From the structure of the M matrix two conditions for

$$A_{yy} = 1 \tag{12.10}$$

follow:

$$M_{1,1/2;1/2} = -M_{-1,1/2;1/2} \tag{12.11}$$

and

$$M_{1,-1/2;1/2} = -M_{-1,-1/2;1/2}. \tag{12.12}$$

Then

$$K_x^{x'} = K_z^{x'} = K_x^{z'} = K_z^{z'} = K_{xy}^{x'} = K_{yz}^{x'} = K_{xy}^{z'} = K_{xz}^{z'} = 0 \tag{12.13}$$

and also

$$C_{x,x} = C_{z,x} = C_{x,z} = C_{z,z} = C_{xy,x} = C_{yz,x} = C_{xy,z} = C_{yz,z} = 0 \tag{12.14}$$

and further:

$$p^{y'} = -A_{0,y} = K_{yy}^{y'} = -C_{yy,y} \tag{12.15}$$

and

$$K_{xx}^{y'} = -C_{xx,y} \tag{12.16}$$

and

$$K_{zz}^{y'} = -C_{zz,y} \tag{12.17}$$

Reactions with the spin structure

$$1^+ + 0^+ \rightarrow 0^+ + 0^- \tag{12.18}$$

have been discussed in Ref. [12]. They have analyzing powers (in the transverse frame) of

$$^T T_{20} = \sqrt{\frac{1}{2}}, \quad \text{i.e. } A_{yy} = 1. \tag{12.19}$$

The reaction

$$^{12}C(\overrightarrow{^{6}Li}, \alpha)^{14}N^{*}(0^{+}, 4.92\,\text{MeV}) \tag{12.20}$$

was used to create an absolute standard for polarized Li beams and secondary standards were derived thereof [13].

Due to its spin structure and symmetry the reaction $^{1}H(\overrightarrow{^{7}Li}, \alpha)\alpha$ has an absolute second-rank tensor analyzing power of $T_{20} = -1$. corresponding to $^{T}T_{20} = 0.5$ in a transverse coordinate system, for a detector at $0°$, and has therefore been used as polarization monitor. Zupranski et al. [12] showed that there exists a relation between the odd-rank tensor moments $^{T}T_{10}(p)$, $^{T}T_{10}(^{7}Li)$, and $^{T}T_{30}(^{7}Li)$ which—under the assumption of $^{T}T_{10}(p) = \pm 1.04$—would allow an absolute determination of $^{T}T_{10}(^{7}Li)$ and $^{T}T_{30}(^{7}Li)$. Similarly, for the reaction $^{23}Na(\overrightarrow{p}, \alpha_{0})^{20}Ne$ which has the same spin structure and for which such a point with maximum vector analyzing power $^{T}T_{10} = 1.05 \pm 0.05$ at $E_{p} = 7.76\,\text{MeV}$ and $\Theta_{lab} = 115°$ was found [14], the possibility of an absolute calibration of all three polarization components of a $^{23}\overrightarrow{Na}$ beam at $E_{lab} = 177\,\text{MeV}$ was suggested.

12.1.4 Calibration Due to Special Conditions

As shown in Chap. 6 analyzing powers of uneven rank, e.g. A_{y} vanish if only a single matrix element contributes to the reaction. To a good approximation this is fulfilled for the two mirror reactions

$$^{3}H(\overrightarrow{d}, n)^{4}He \tag{12.21}$$

and

$$^{3}He(\overrightarrow{d}, p)^{4}He \tag{12.22}$$

at very low energies, e.g. around the $J^{\pi} = 3/2^{+}$ resonance at $E_{lab} = 107\,\text{keV}$ and $430\,\text{keV}$, respectively. Tensor moments of even rank (here: the tensor analyzing powers) will not vanish, but are calculable in addition, as long as no admixtures of other (generally unknown) matrix elements have to be considered. With the formalism of Chap. 6 one obtains:

$$A_{zz} = -P_{2}(\cos\Theta), \quad A_{xz} = -\frac{1}{2}P_{2}^{1}(\cos\Theta), \quad A_{xx} - A_{yy} = -\frac{1}{2}P_{2}^{2}(\cos\Theta). \tag{12.23}$$

Thus these reactions are good analyzing reactions for deuteron tensor polarization as long as assumed admixtures of possible $1/2^{+}$ S wave and higher-waves matrix elements (the contributions of which probably are on the order of 1%) may be neglected.

Comparable analyzing reactions for the vector polarization of deuterons as well as protons do not exist at low energies because these reactions are normally dominated by S waves. The vector analyzing power, in order to be non-zero, requires an interference at least with P waves and therefore will be quite small. In this respect the ^{2}H($\vec{\text{d}}$, n)^{3}He and ^{2}H($\vec{\text{d}}$, p)^{3}H reactions are exceptional showing A_y of about 0.2 even at $E_{lab} < 30$ keV.

12.1.5 Typical Low-Energy Analyzer Reactions

The requirements on analyzing reactions are: high cross-section, high analyzing power, smooth, slowly-varying behavior of both with energy and angles, and easily available and stable target material. Thus, for general use with polarized beams from low-energy accelerators a few standard analyzing reactions have evolved and have been calibrated in careful measurements. These are:

- For protons:

  - ^{4}He($\vec{\text{p}}$, p)^{4}He
  - ^{12}C($\vec{\text{p}}$, p)^{12}C
  - ^{28}Si($\vec{\text{p}}$, p)^{28}Si or ^{nat}Si($\vec{\text{p}}$, p)^{nat}Si

- For deuterons:

  - ^{4}He($\vec{\text{d}}$, d)^{4}He
  - ^{3}He($\vec{\text{d}}$, p)^{4}He
  - ^{3}H($\vec{\text{d}}$, n)^{4}He
  - ^{2}H($\vec{\text{d}}$, p)^{3}H and ^{2}H($\vec{\text{d}}$, n)^{3}He

Only two of these reactions will be discussed here in more detail, for the others the reader is referred to the literature cited.

12.1.5.1 The ^{4}He($\vec{\text{p}}$, p)^{4}He Reaction for Protons

For the measurement of the (proton or deuteron) vector polarization it is sufficient to measure a left-right asymmetry, e.g. with two detectors at some polar angle Θ with azimuthal angles Φ separated by 180° (depending on the definition of the coordinate system) 0°, 180°, or $-90°$, $+90°$. Since parity conservation allows only the component A_y with $\hat{y} = \frac{\vec{k}_{in} \times \vec{k}_{out}}{|k_{in} \times k_{out}|}$ to be non-zero, "left/right" refers to this polarization/analyzing power component. The ^{4}He($\vec{\text{p}}$, p)^{4}He elastic scattering has been studied extensively (see e g. [15]), but for polarimeter purposes it is sufficient to recognize that it has at least one absolute calibration point at $E_p = 12.1$ MeV and $\Theta_{lab} = 112°$. see Sect. 12.1.2. With this constant angle setting it is useful to have an excitation function over the relevant energy range. Such an excitation function was

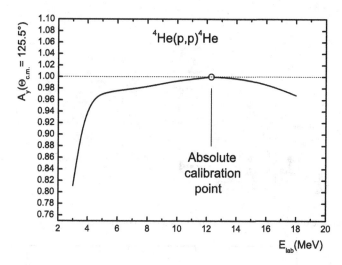

Fig. 12.1 Excitation function of the reaction ^{4}He($\vec{\text{p}}$, p)^{4}He at $\Theta_{c.m.} = 125.5°$ with the absolute calibration point at $E_{lab} = 12.1$ MeV

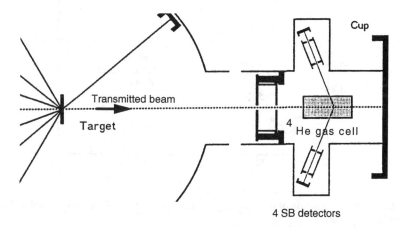

Fig. 12.2 Proton polarimeter using the ^{4}He($\vec{\text{p}}$, p)^{4}He reaction behind the scattering chamber [16]

taken from the more comprehensive measurement of [15] and is shown in Fig. 12.1. An example of a proton polarimeter behind a scattering chamber is shown in Fig. 12.2.

12.1.5.2 The ^{3}He($\vec{\text{d}}$, p)^{4}He Reaction for Deuterons

This reaction is, up to about $E_d = 450$ keV. dominated by the $J^\pi = 3/2^+$ S-wave resonance and is therefore not sensitive to deuteron vector polarization, whereas

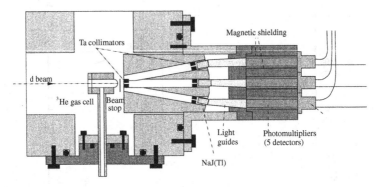

Fig. 12.3 Scheme of the Cologne ^{3}He$(\vec{d}, p)^4$He deuteron polarimeter

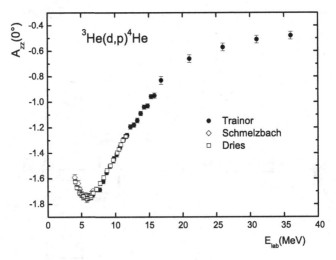

Fig. 12.4 Excitation function of the analyzing power $A_{zz}(0°)$ of the ^{3}He(d,p)^{4}He reaction. Data are from Refs. [20–22]

the tensor analyzing power is large and well-known. At higher energies all four components A_y, A_{zz}, A_{xx-yy}, and A_{xz} are $\neq 0$ and have been calibrated. A typical vector- and tensor polarimeter based on this reaction consists of five detectors, i.e. one central detector at $\Theta = 0°$ and four detectors at $\Theta = 24.5°$, $\Phi = 0°$, 90°, 180°, and 270° in a 4π arrangement. Figure 12.3 shows the design of such a polarimeter [17]. In addition, the published and newer unpublished values of the analyzing powers are shown in Figs. 12.4 and 12.5. It is evident that the quality of some data is not very good, especially concerning systematic errrors.

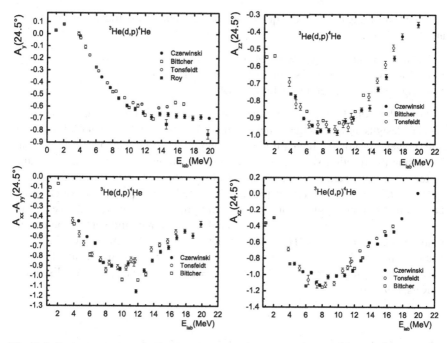

Fig. 12.5 Excitation functions at $\Theta_{lab} = 24.5°$ of analyzing powers of the ^{3}He(d, p)^{4}He reaction. Data are from Refs. [18, 23, 24]

12.1.5.3 Deuteron Polarimetry at Very Low (Astrophysical) Energies

The fact that important analyzing reactions such as the ^{3}H($\vec{d}$,n)^{4}He and the ^{3}He($\vec{d}$,p)^{4}He reactions basically proceed via one (resonant) matrix element makes them sensitive to tensor polarization, but insensitive to deuteron vector polarization. Good analyzing reactions for all polarization components down to <20 keV are the ^{2}H($\vec{d}$,n)^{3}He and ^{2}H($\vec{d}$,p)^{3}H reactions which show surprisingly large vector analyzing powers. This is mainly due to large P (and even D) wave components in the nuclear interaction. Good data exist from several authors (for a survey see e.g. Ref. [19]).

12.1.6 Polarimetry in Polarization-Transfer Experiments

Here polarization components of the ejectiles from a reaction have to be determined as functions of the polarization of the beam incident on the first-reaction target. The low intensity of the incoming particles requires special designs for the polarimeters and compromises with respect to attainable precision in the available beam-time. This involves some averaging over energies and angles and calibration of these averaged analyzing powers. The quantities characterizing these transfer polarimeters are

- the "effective analyzing powers $\langle A_y \rangle$,
- the effective cross-sections and relative "efficiencies" ϵ,
- the figures of merit defined as f.o.m. $= \langle A \rangle^2 \cdot \epsilon$.

Depending on the type of particles and of the polarization components some special designs have emerged such as the vane or "venetian-blind" design for polarimeters in order to increase the efficiency by limiting the angular spread but increasing the reaction volume. The reactions which have been used are the same as for beam-polarization measurements, as listed above. Details can be found e.g. in Refs. [25, 26, 28]. References [26–28] presented a modular design which could be adapted to vector polarization of outgoing protons or deuterons as well as to deuteron tensor polarization (which was applied to the measurement of the elastic scattering reaction ^{2}H($\vec{p}$, $\vec{d}$)^{1}H [28]. Polarimeters optimized for measuring the polarization transfer of the ^{2}H($\vec{d}$, $\vec{p}$)^{3}H reaction, based on Si as analyzer target, at the very low incident deuteron energies of $E_{lab} = 90$ and 58 keV have been discussed in Refs. [29, 30].

12.2 Polarimetry of Atomic (and Molecular) Beams

In storage rings (colliders etc.) increasingly polarized atomic beams are used as targets, generally feeding into storage cells, are used as targets (e.g. at HERMES/ DESY [31], IUCF, EDDA/COSY and ANKE/COSY). During the development (but also during their use) it appears very useful to be able to measure the polarization directly and absolutely without use of special calibrated low-energy nuclear reactions which may also be difficult to apply (e.g. due to very low vector analyzing power at these energies or because of requiring tritium as a target). Several different schemes have been developed:

- Breit–Rabi polarimeters
- Lambshift ("spinfilter") polarimeters.

12.2.1 Breit–Rabi Polarimeters

Like in optical polarimetry where polarizers and polarization analyzers are similar devices the principle of these polarimeters is a Stern–Gerlach apparatus with radiofrequency transitions. Figure 12.6 shows schematically a typical setup similar to the one used at the DESY-HERMES experiment, e. g. [32]. The occupation of each single Zeeman hyperfine state is determined by the system of spin-separation (multipole) magnets combined with RF transition units such that all but one state is being

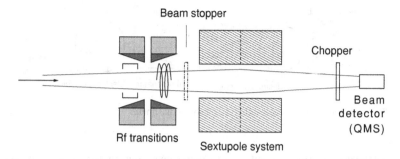

Fig. 12.6 Typical setup of a Breit–Rabi polarimeter for atomic beams of hydrogen or deuterium. Different numbers of RF-transition units and of sextupoles have been used

transmitted. The measured intensities together with transition probabilities for a number of different transitions form a set of coupled equations which can be solved resulting in occupation numbers from which polarizations are evaluated. Very good vacuum throughout the system is needed in order not to contaminate the beam with unpolarized residual gas atoms. Problems arise if the spin-state separation is not 100%, and when the RF transition probability is less than 100%.

The sextupoles focus the $m_J = +1/2$ component onto the beam detector while defocussing the $m_J = -1/2$ component. Without RF transitions the signal in the detector is

$$S_0 = c \sum_a \sigma_a I_a \tag{12.24}$$

with c a calibration factor of the detection system (which includes the detection efficiency and geometrical acceptance etc.), σ_a the probability of the atoms in the hyperfine state $|a\rangle$ to be transmitted by the sextupole system into the detector, and I_a the intensity of the atoms in the hyperfine state $|a\rangle$ in the incident beam. Any RF transition in the RF transition units will redistribute the occupation numbers among the possible hyperfine states. Then the signal at the detector will be

$$S_i = c \sum_a \left(\sum_b \sigma_b T_{ba}^i \right) I_a = c \sum_a M_{ia} I_a \tag{12.25}$$

where T_{ba}^i is a matrix describing the occupation exchange between different hyperfine states by the transition i. It contains the matrix M_{ia} which is a linear function of the transition efficiencies ϵ_{ia} times the transition probabilities σ_i between all states involved. The linear system of equations can be solved if a sufficient number of separate measurements with different transitions can be performed. Four (six) such measurements are required for solutions by matrix inversion. However, the errors of the different quantities involved may require more than the minimal number of measurements leading to a problem of least-squares minimization. The resulting

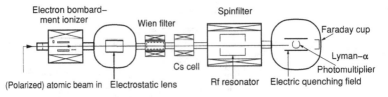

Fig. 12.7 Schematic of the Cologne-Jülich Lambshift polarimeter

errrors of the polarizations are obtained from the covariance matrix of the polarization vector. The authors of [33] claim an absolute precision of the measured polarizations of about 0.01 and a sensitivity of better than 10^{13} atoms/s.

12.2.2 Lambshift Polarimeters LSP

Figure 12.7 shows the general setup of a typical Lambshift polarimeter with a spin filter and a Glavish-type electron bombardment ionizer for the polarization measurement on atomic H or D beams [34]. This scheme uses the Lambshift via a spin filter as central element [35], (see Sect. 8.5 on the Lambshift polarized-ion source) which selects and transmits single Zeeman components of the 2S hyperfine structure by the combined action of a static magnetic field, a static electric field (Stark quenching) and a radiofrequency transition field at 1.609 GHz. By varying the resonance condition by scanning the magnetic field all HFS components can be measured. An electric quenching field after the spinfilter quenches the atoms to the ground (1S) state thereby emitting Lyman-α radiation at 121.5 nm which is registered by a sensitive and selective photomultiplier. The intensities in each such Lyman-α peak produce the polarization values of the beam (vector as well as tensor). By applying a number of corrections it is possible to get the absolute polarization of the beam because all these corrections are known and can be calculated exactly [34]

Because the LSP for H^+/D^+ ions works very much like the LSS described above, the beams should have energies of about 500 or 1000 eV, respectively, then undergo a charge exchange into metastables. A neutral atomic beam first has to be ionized efficiently, e.g. by electron impact or ECR ionizers, accelerated and then transmuted into metastables by Cs vapor. The possibility of measuring the polarization of negative ions H^- or D^- after a double charge exchange (e.g. in ^{16}O or even plain air, or in ^4He [36, 37]) converting them to positive ions, before the charge exchange into metastables, is presently being investigated (Engels, R.: private communication (2010)). The sensitivity in determining polarization absolutely is such that measurements with beams of as low as 10^{14} particles/s in less than 1 min, especially when using lock-in techniques, are possible. The LSP also has the potential of achieving higher sensitivity e.g. by improving the Lyman-α collection efficiency and other measures.

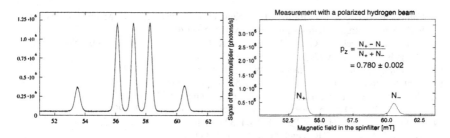

Fig. 12.8 Ly-α spectra of mixed unpolarized H and D beams (*left*) and polarized H (*right*)

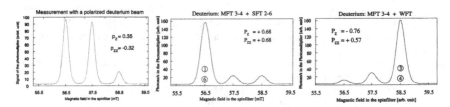

Fig. 12.9 Ly-α spectra of D with different polarizations

A number of LSP have been implemented at different laboratories and were used mainly for polarized-ion source development especially to optimize the efficiency of RF transitions, but also for beam/target polarization control [38–41]. Figures 12.8 and 12.9 show Ly-α spectra obtained with unpolarized and differently prepared polarized H and D beams.

References

1. Luck, W., Ph.d. thesis, Universität Marburg, unpublished (1989)
2. Timsit, R.S., Daniels, J.M.: Can J. Phys. **49**, 545 (1971)
3. Leemann, Ch., Bürgisser, H., Huber, P., Rohrer, U., Paetz gen. Schieck, H., Seiler, F.: Helv. Phys. Acta **44**, 141 (1971)
4. Huber, P., Leemann, Ch., Rohrer, U., Seiler, F.: Helv. Phys. Acta. **42**, 907 (1969)
5. Plattner, R., Bacher, A.D.: Phys. Lett. **36B**, 211 (1971)
6. Seiler, F.: Phys. Lett. **61B**, 144 (1976)
7. Jacobsohn, B.A., Ryndin, R.M.: Nucl. Phys. **24**, 505 (1961)
8. Darden, S.E., Prior, R.M., Corrigan, K.W.: Phys. Rev. Lett. **25**, 1673 (1970)
9. König, V., Grüebler, W., Ruh, A., White, R.E., Schmelzbach, P.A., Risler, R., Marmier, P.: Nucl. Phys. **A166**, 393 (1971)
10. Keaton, P. W. Jr., Armstrong, D.D., Lawrence, G.P., McKibben, J.L., Ohlsen, G.G.: In: Proceedings of the 3rd International Symposium on Polarization Phenomena in Nuclear Reactions, Madison 1970, p. 849. University of Wisconsin Press, Madison (1971)
11. Sagara, K.: Few-Body Systems **48**, 59 (2010)
12. Zupranski, P., Dreves, W., Egelhof, P., Steffens, E., Fick, D., Rösel, F.: Nucl. Instrum. Methods **167**, 193 (1979)

13. Cathers, P.D., Green, P.V., Bartosz, E.E., Kemper, K.W., Marechal, F., Myers, E.G., Schmidt, B.G.: Nucl. Instrum. Methods Phys. Res. A **457**, 509 (2001)
14. Paetz gen. Schieck, H., Gaiser, N.O., Nyga, K.R., Prior, R.M., Darden, S.E.: Nucl. Instrum. Methods Phys. Res. A **25**, 616 (1987)
15. Schwandt, P., Clegg, T.B., Haeberli, W.: Nucl. Phys. **A163**, 432 (1971)
16. Vohl, S.: Diploma thesis, Universität zu Köln, unpublished (1990)
17. Engels, R.: Diploma thesis, Universität zu Köln, unpublished (1997)
18. Czerwinski, A.: Diploma thesis, Universität zu Köln, unpublished (1999)
19. Paetz gen. Schieck, H.: Eur. Phys. J. A **44**, 321 (2010)
20. Trainor, T.A., Clegg, T.B.: Nucl. Phys. **A220**, 533 (1974)
21. Schmelzbach, P.A., Grüebler, W., König, V., Risler, R., Boerma, D.O., Jenny, B.: Nucl. Phys. **A264**, 45 (1976)
22. Dries, L.J., Clark, H.W., Detomo, R., Regner, J., Donoghue, T.R.: Phys. Rev. C **21**, 475 (1980)
23. Tonsfeldt, S.A.: Dissertation, University of North Carolina at Chapel Hill, (1980). Available from University Microfilms International, A Bell & Howell Comp., Ann Arbor, Michigan, USA
24. Bittcher, M., Grüebler, W., König, V., Schmelzbach, P.A., Vuaridel, B., Ulbricht, J.: Few-Body Systems. **9**, 165 (1990)
25. Ohlsen, G.G., Keaton, P.W.Jr.: Nucl. Instrum. Methods **109**, 41 and 61 (1973)
26. Vohl, S.: Ph.D. thesis, Universität zu Köln, unpublished (1995)
27. Sydow, L., Vohl, S., Lemaître, S., Nießen, P., Nyga, K.R., Reckenfelderbäumer, R., Rauprich, G., Paetz gen. Schieck, H.: Nucl. Instrum. Methods Phys. Res. A **327**, 441 (1993)
28. Sydow, L., Vohl, S., Lemaître, S., Patberg, H., Reckenfelderbäumer, R., Paetz gen. Schieck, H., Glöckle, W., Hüber, D., Witała, H.: Few-Body Systems. **25**, 133 (1998)
29. Katabuchi, T., Kudo, K., Masuno, K., Iizuka, T., Aoki, Y., Tagishi, Y.: Phys. Rev. C **64**, 047601 (2001)
30. Imig, A., Düweke, C., Emmerich, R., Ley, J., Zell, K.O., Paetz gen. Schieck, H.: Phys. Rev. C **73**, 024001 (2006)
31. Airapetian, A. et al.: HERMES Collaboration. Nucl. Instrum. Methods Phys. Res. A **540**, 68 (2005)
32. Baumgarten, C. et al.: Nucl. Instrum. Methods Phys. Res. A **496**, 263 and 277 (2003)
33. Baumgarten, C. et al.: Nucl. Instrum. Methods Phys. Res. A **482**, 606 (2002)
34. Engels, R., Emmerich, R., Ley, J., Tenckhoff, G., Paetz gen. Schieck, H.: Rev. Sci. Instrum. **74**, 4607 (2003)
35. Ohlsen, G.G., McKibben, J.L., Lawrence, G.P., Keaton, P.W.Jr., Armstrong, D.D.: Phys. Rev. Lett. **27**, 599 (1971)
36. Tawara, H., Russek, A.: Rev. Mod. Phys. **45**, 178 (1973)
37. Tawara, H., Russek, A.: At. Nucl. Data Tables **22**, 491 (1978)
38. Lemieux, S.K., Clegg, T.B., Karwowski, H.J., Thompson, W.J.: Nucl. Instrum. Methods Phys. Res. A **333**, 434 (1993)
39. Emmerich, R.: Ph. D. thesis, Universität zu Köln (2007). http://kups.ub.uni-koeln.de/volltexte/2007/2081
40. Emmerich, R., Paetz gen. Schieck, H.: Nucl. Instrum. Methods Phys. Res. A **586**, 387 (2008)
41. Kremers, H.R., Beijers, J.P.M., Kalantar-Nayestanaki, N., Clegg, T.B.: Nucl. Instrum. Methods Phys. Res. A **516**, 209 (2004)

Part V
Applications

Chapter 13
Medical Applications

13.1 Hyperpolarized ^{3}He and ^{129}Xe

The recent progress in medical imaging techniques such as magnetic-resonance imaging (*nmr* or *mri*), computer tomography (CT with X-rays), and positron-emission tomography (PET scanning using short-lived radioactive nuclei) has been impressive. Two areas where diagnostic tools lacked behind have been tomography of the blood vessels of the brain and of the bronchi. Starting around 1995 new ideas of imaging these have been realized: use of hyperpolarized ^{3}He and ^{129}Xe. Very recently also the metabolism of compounds containing other nuclei with spins $\neq 0$ such as ^{13}C,^{2}H, ... has been studied, making use of the much improved *nmr* signal strength of hyperpolarized specimens which in these cases have been polarized by brute force (low temperatures at very high magnetic fields) combined with RF transitions in solids (Overhauser effect). An example of such projects is "Nuclear Spin Imaging" (NSI), i.e. *mri* with brute-force hyperpolarized spin-1/2 nuclei such as ^{3}He,^{13}C,^{15}N,^{19}F, and ^{31}P using magnetic fields of B $\approx$ 17 T and T $\approx$ 10 mK, see [1].

Considering the signal strength in conventional nuclear magnetic resonance *nmr/mri* we see that it is governed by the density of resonant nuclei (such as the protons in water) and by the very small "brute-force" polarization in strong magnetic fields and (normally) room temperature. Figure 13.1 shows the situation for an $S = 1/2$ spin system. Assuming a Boltzmann distribution the occupation numbers of the spin-1/2 system are

$$N_{\pm} = N_0 \exp\left(\mp \frac{1}{2}\gamma\hbar\frac{B}{kT}\right) \tag{13.1}$$

where $\gamma = \mu_N g_N/\hbar$ are the gyromagnetic ratio, μ_N the nuclear magneton, g_N the nuclear g factor, and the transition frequency between the two substates is $\omega = \gamma B$. Thus the "brute-force" nuclear polarization is

H. Paetz gen. Schieck, *Nuclear Physics with Polarized Particles*,
Lecture Notes in Physics 842, DOI: 10.1007/978-3-642-24226-7_13,
© Springer-Verlag Berlin Heidelberg 2012

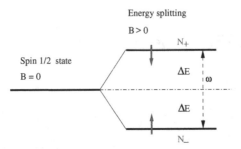

Fig. 13.1 Schematic of energy splitting $2 \cdot \Delta E = \gamma \hbar B$, occupation numbers N_+ and N_- of substates and transition frequency ω of a spin-1/2 system in a magnetic field B.

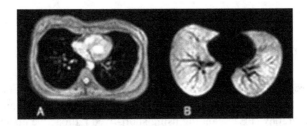

Fig. 13.2 Comparison of *mri* images of the chest of a healthy person using conventional proton imaging (*left*) and hyperpolarized ^{3}He (*right*). The complementary function of both methods is clearly visible. The figure is from [2], courtesy of American College of Chest Physicians.

$$p = \frac{N_+ - N_-}{N_+ + N_-} = \tanh\left(\frac{1}{2}\gamma\hbar\frac{B}{kT}\right) \approx \frac{1}{2}\gamma\hbar\frac{B}{kT} \qquad (13.2)$$

which is normally very small and reaches substantially higher values only at very high magnetic fields and very low temperatures. Even at $B = 1$ T and room temperature the polarization of hydrogen nuclei is only about $3 \cdot 10^{-6}$. If we take a hyperpolarized gas such as ^{3}He with a density about a thousand times smaller than water, but polarized to 60% then the signal strength is still many times higher than with *mri* in water.

$$\frac{n(^3\text{He}) \cdot p(^3\text{He})}{n(\text{H}_2\text{O}) \cdot p(\text{proton})} \approx 10^4 \qquad (13.3)$$

The clinical use of hyperpolarized ^{3}He consists of letting the patient inhale the gas and keep his breath for about a minute while an *mri* scan is taken. The *mri* apparatus is not very different from conventional *mri* on protons, except that the transition frequency per 1 T is 33.2 MHz instead of 42.5 MHz. Contrary to proton *mri* the lung itself does not produce an image but the hollow spaces of the bronchi do and show possible anomalies such as tumors, emphysema, chronic obstructive pulmonary disease (COPD), etc. Figure 13.2 compares both methods and shows their complementarity. In a number of places/universities worldwide (e.g. Princeton, Duke, Virginia, Mainz) the collaboration between nuclear and atomic physicists with medical departments has led to a number of medical installations where lung and brain studies as well as diagnosis are performed. Due to long relaxation times the polarized ^{3}He produced at some physics laboratories is transported to quite distant places while the sample is kept in a homogeneous magnetic holding field.

References

1. Tanaka, M., et al.: In: [3]
2. De Lange, E.E., Altes, T.A., Patrie, J.T., Gaare, J.D., Knake, J.J., Mugler, J.P. III, Platts-Mills, T.A.: Chest **130**, 1055 (2010)
3. Rathmann, F., Ströher, H. (eds.) Proceedings of International Spin Conference (SPIN2010), Jülich 2010. Published as Open Access by IOP Conference series (2011)

Chapter 14
"Polarized" Fusion

Increasing energy demand in view of limited supply, as well as environmental and nuclear-safety concerns leading to increased emphasis on renewable energy sources such as solar or wind energy are expected to focus public and scientific interest increasingly also on fusion energy. With the decision to build ITER (low-density magnetic confinement) and also continuing research on (high-density) inertial-confinement fusion (cf. the inauguration of the laser fusion facility at the Lawrence Livermore National Laboratory) prospects of fusion energy have probably entered a new era. The idea of "polarized fusion", i.e. using spin-polarized particles as nuclear fuel was developed long ago ([1, 2], and for more recent developments see [3, 4]). It offers a number of modifications as compared to conventional unpolarized fusion. The main features are:

- Neutron management: replacement or reduction of neutron-producing reactions in favor of charged-particle reactions.
- Handling of the emission direction of reaction products.
- Increase of the reaction rate.

Some of these improvements may lead to lower ignition limits and to more economical running conditions of a fusion reactor due to less radiation damage and activation to structures and especially the blanket, necessary to convert the neutron energy to heat, or may lead to concepts of a much simpler and longer-lasting blanket. At the same time its realization will meet additional difficulties for which solutions have to be studied. Some of these are:

- Preparation of the polarized fuel, either in the form of intense beams of polarized ^{3}H, D, or ^{3}He atoms or as pellets filled with polarized liquid or solid.
- Injection of the polarized fuel.
- Depolarization during injection or during ignition.

As an example of a recent effort to address some of these questions we cite Refs. [5, 6]. The energy range in which the relevant fusion reactions will take place is <100 keV where the Coulomb barrier strongly suppresses charged-particle cross-sections. This is the reason why necessary experimental polarization data with

H. Paetz gen. Schieck, *Nuclear Physics with Polarized Particles*,
Lecture Notes in Physics 842, DOI: 10.1007/978-3-642-24226-7_14,
© Springer-Verlag Berlin Heidelberg 2012

sufficiently high precision such as spin-correlated cross-sections have not been measured. Existing reaction analyses and predictions for polarized fusion relied on existing world data sets of other (simpler) data. On the other hand, sufficiently micro-scopic and therefore realistic theoretical predictions (such as for the three-nucleon system) are just beginning to become available for the four-nucleon systems at the required low energies [7]. An interesting question is whether the recently discussed electron-screening enhancement ([8] and references therein) of the very-low energy cross-sections has any bearing on polarized fusion.

It should be mentioned here that in the past polarization observables played a deci-sive role in elucidating the reaction mechanisms of few-body reactions as well as the nuclear structure of few-body nuclei, especially in the two- to six-body systems. At present only four- or five-nucleon systems are considered for fusion energy. The quan-tities relevant for fusion-energy studies are the integrated (or total) cross-section σ, the reaction coefficient (or reaction parameter) $\langle \sigma v \rangle$, and the (relative) power density P_f.

14.1 Five-Nucleon Fusion Reactions

The important reactions to be discussed here are:

- $d + {}^3H \rightarrow n + {}^4He + 17.58\,\mathrm{MeV}$.
- $d + {}^3He \rightarrow p + {}^4He + 18.34\,\mathrm{MeV}$.

The two mirror reactions have some very pronounced features: At the low energies discussed here both proceed via strong S-wave resonances (at deuteron lab. energies of 107 keV for ${}^3H(d,n){}^4He$, and 430 keV for ${}^3He(d,p){}^4He$, respectively). These reso-nant states are quite pure $J^\pi = 3/2^+$ states with possibly very little admixture of a $J^\pi = 1/2^+$ S-wave and/or higher-wave contributions. This has been a long-time point of discussion, mainly because of the reactions being very good absolute tensor-polarization analyzers, provided they proceed only through the S-wave, $J^\pi = 3/2^+$ state. Experimental evidence shows that other contributions are small (of the order of a few %). An example of the ${}^3He(d,p){}^3He$ reaction on resonance is an early spin-correlation measurement [9, 10] supporting this assumption, see Fig. 14.1. For a recent discussion of this reaction at low energies see e.g. Refs. [11, 12]. The results for the mirror reaction ${}^3H(d,n){}^4He$ are similar.

The relatively good knowledge about these two reactions allows the conclusion that with polarized beams and targets an enhancement of the fusion yield close to a factor of 1.5 may be expected. A simple hand-waving statistical argument shows that the reactions, if they go through the $3/2^+$ state and with the entrance channel prepared in a stretched configuration, as compared to the unpolarized entrance channel with a purely statistical spin configuration, yields just this enhancement.

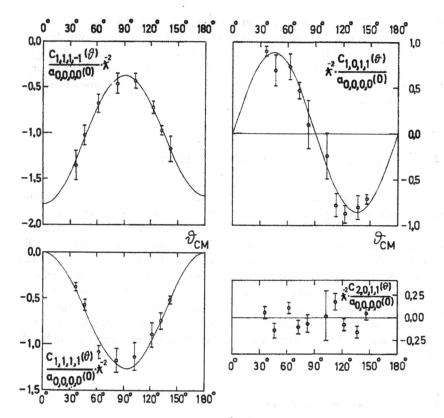

Fig. 14.1 Spin-correlation measurement of the $^3\vec{\text{He}}(\vec{\text{d}}, \text{p})^4\text{H}$ reaction at $E_d(\text{lab}) = 430\,\text{keV}$. This energy corresponds to that of the S-wave $J^\pi = 3/2^+$ resonance. By permission of Birkhäuser Verlag, Basel

14.2 Four-Nucleon Reactions

The most important four-nucleon fusion reactions are the D + D reactions which in a plasma also inevitably accompany the more important five-nucleon reactions discussed above:

- $d + d \rightarrow n + {}^3\text{He} + 3.268\,\text{MeV}$.
- $d + d \rightarrow p + {}^3\text{H} + 4.033\,\text{MeV}$.

Whereas the situation of the five-nucleon systems is relatively clear-cut the four-nucleon systems and especially the two DD reactions have a number of problems in their description, especially in view of "polarized fusion". Different from the five-nucleon case the non-resonant reaction mechanism is very complicated (at least 16 complex matrix elements including S-, P-, and D-waves have to be considered with spin-flip transitions from the entrance to the exit channel which contribute even at low energies). One consequence of participating P waves is that they are the only reactions with appreciable vector- (besides tensor-)analyzing power even down to 20 keV lab. energy which makes them very useful analyzer reactions at these energies (see also Sect. 12.1.5). In a semi-classical picture this is made plausible with the large extension of the deuteron wave function and therefore large interaction distance of the two deuterons.

14.2.1 Suppression of Unwanted DD Neutrons

Aneutronic fusion may have a number of advantages (not the least unimportant *economic* ones) over the use of neutron-producing reactions. At an advanced stage the ^{3}H(d,n) reaction could be replaced by the ^{3}He(d,p) reaction. However, DD neutrons would remain. It has been suggested by theoretical approaches that DD neutrons could be reduced substantially by polarizing the deuterons, thus forming a quintet ($S = 2$) state. The main argument was that quintet states in the entrance channel would require spin-flip transitions which are Pauli-forbidden in first order. However, this argument would be invalid if the reactions proceeded via the D state of the deuteron, and so far the (indirect) experimental evidence does not support this conjecture, see e.g., [4]. A direct spin-correlated cross-section measurement is still lacking, but is highly desirable.

14.2.1.1 Evidence for Suppression?

Lacking a direct spin-correlation experiment at very low energies, two indirect approaches have been taken.

- Parametrization of world data by a multi-channel R-matrix analysis [13].
- Köln parametrization of world data of the ^{2}H(d,n)^{3}He and ^{2}H(d,p)^{3}H reactions by direct T-matrix analysis below 1.5 MeV [14–18].

Both approaches allow predictions observable of the DD reactions, also of the *quintet suppression factor QSF*, as defined below.

14.2.1.2 Definition of QSF

In order to quantify the extent to which DD neutrons may be suppressed by polarizing the fusion fuel nuclei the "Quintet Suppression Factor (QSF)" is defined as:

$$QSF = \frac{\sigma_{1,1}}{\sigma_0} \tag{14.1}$$

where

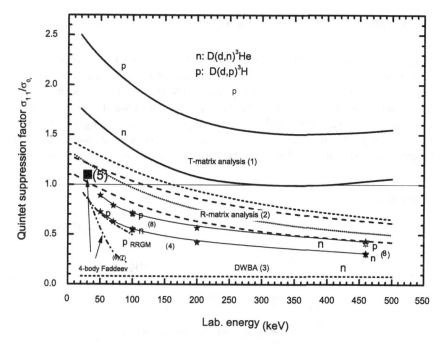

Fig. 14.2 Quintet suppression factor as predicted by various theoretical and from two experimental approaches using world data of DD reactions. The relevant references (numbers in parentheses in the figure) are: (1) [16], (2) [20, 21], (3) [22, 23], (4) [24–26], (5) [27], (6) [28], (7) [29], and (8) [7, 19]. The predictions from Refs. [7, 19, 27, 28, 26] are from microscopic Faddeev–Yakubovsky calculations

$$\sigma_0 = \frac{1}{9}(\underbrace{2\sigma_{1,1}}_{\text{Quintet}} + \underbrace{4\sigma_{1,0}}_{\text{Triplet}} + \underbrace{\sigma_{0,0} + 2\sigma_{1,-1}}_{\text{Singlet}}) \qquad (14.2)$$

is the total (integrated) cross-section to which the four independent channel-spin cross-sections $\sigma_{1,1}$ (spin-quintet configuration), $\sigma_{1,0}$ (spin triplet), $\sigma_{0,0}$, and $\sigma_{1,-1}$ (two spin-singlet terms) contribute with their statistical weights.

In Fig. 14.2 all results for the QSF from different theoretical predictions as well as from the two data parametrizations for both DD reactions are shown.

The theoretical approaches reach from DWBA calculations to—very recently— microscopic calculations including thr Coulomb force [7, 19] and vary widely. However, these latest calculations are the most advanced ones and lend confidence to the idea that substantial suppression occurs only in the higher energy range, i.e., above the region of the Gamow peak where fusion-energy production will take place.

14.3 Status of "Polarized" Fusion

In view of the wide range of theoretical predictions and the lack of direct experimental evidence e.g., for the QSF it seems mandatory to perform a direct D + D spin-correlation experiment in the energy range from 10 to 100 keV. The number of correlation coefficients, however, is quite formidable. The general cross-section for the reaction of a spin-1 polarized beam with a polarized spin-1 target contains—besides the unpolarized cross-section—analyzing powers of beam and of target in addition to the 32 spin-correlation terms. Parity conservation has been taken into account.

$$[\sigma(\Theta, \Phi)]_{\Phi=0} =$$

$$\sigma_0(\Theta)\left\{ 1 + \frac{3}{2}\left[A_y^{(b)}(\Theta)p_y + A_y^{(t)}q_y \right] + \frac{1}{2}\left[A_{zz}^{(b)}(\Theta)p_{zz} + A_{zz}^{(t)}(\Theta)q_{zz} \right] \right.$$

$$+ \frac{1}{6}\left[A_{xx-yy}^{(b)}(\Theta)p_{xx-yy} + A_{xx-yy}^{(t)}(\Theta)q_{xx-yy} \right]$$

$$+ \frac{2}{3}\left[A_{xz}^{(b)}(\Theta)p_{xz} + A_{xz}^{(t)}(\Theta)q_{xz} \right]$$

$$+ \frac{9}{4}\left[C_{y,y}(\Theta)p_y q_y + C_{x,x}(\Theta)p_x q_x + C_{x,z}(\Theta)p_x q_z \right.$$

$$\left. + C_{z,x}(\Theta)p_z q_x + C_{z,z}(\Theta)p_z q_z \right]$$

$$+ \frac{3}{4}\left[C_{y,zz}(\Theta)p_y q_{zz} + C_{zz,y}(\Theta)p_{zz}q_y \right]$$

$$+ \quad C_{y,xz}(\Theta)p_y q_{xz} + C_{xz,y}(\Theta)p_{xz}q_y + C_{x,yz}(\Theta)p_x q_{yz}$$

$$+ \quad C_{yz,x}(\Theta)p_{yz}q_x + C_{z,yz}(\Theta)p_z q_{yz} + C_{yz,z}(\Theta)p_{yz}q_z$$

$$+ \frac{1}{4}\left[C_{y,xx-yy}(\Theta)p_y q_{xx-yy} + C_{xx-yy,y}(\Theta)p_{xx-yy}q_y \right.$$

$$\left. + C_{zz,zz}(\Theta)p_{zz}q_{zz} \right]$$

$$+ \frac{1}{3}\left[C_{zz,xz}(\Theta)p_{zz}q_{xz} + C_{xz,zz}(\Theta)p_{xz}q_{zz} \right]$$

$$+ \frac{1}{12}\left[C_{zz,xx-yy}(\Theta)p_{zz}q_{xx-yy} + C_{xx-yy,zz}(\Theta)p_{xx-yy}q_{zz} \right]$$

$$+ \frac{4}{9}\left[C_{xz,xz}(\Theta)p_{xz}q_{xz} + C_{yz,yz}(\Theta)p_{yz}q_{yz} \right]$$

$$+ \frac{8}{9}\left[C_{xy,yz}(\Theta)p_{xy}q_{yz} + C_{yz,xy}(\Theta)p_{yz}q_{xy} \right]$$

$$+ \frac{16}{9}C_{xy,xy}(\Theta)p_{xy}q_{xy}$$

$$+ \frac{1}{9}\left[C_{xz,xx-yy}(\Theta)p_{xz}q_{xx-yy} + C_{xx-yy,xz}(\Theta)p_{xx-yy}q_{xz} \right]$$

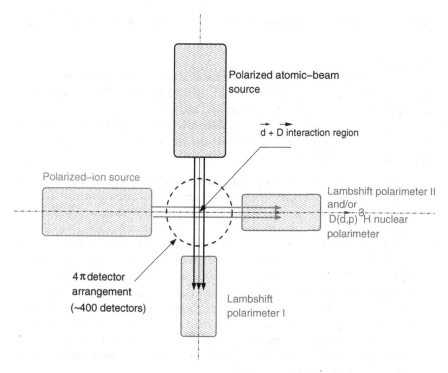

Fig. 14.3 Scheme of a possible spin-correlation experiment with an atomic $\vec{D}$ beam crossing a $\vec{d}$ ion beam of 10–100 keV. A granular 4π system of detectors surrounds the interaction region

$$+ \frac{1}{36} C_{xx-yy,xx-yy}(\Theta)\, p_{xx-yy} q_{xx-yy}$$

$$+ \frac{1}{2}\left[C_{x,xy}(\Theta)\, p_x q_{xy} + C_{xy,x}(\Theta)\, p_{xy} q_x + C_{z,xy}(\Theta)\, p_z q_{xy} \right.$$

$$\left. + C_{xy,z}(\Theta)\, p_{xy} q_z \right] \Big\} \tag{14.3}$$

In the case of identical particles with spin 1 such as deuterons a few terms in this formula are redundant because exchange of beam and target nuclei leads to simple relations (identities, when related to the same coordinate system) between a number of relevant correlation coefficients, thereby reducing the number of independent coefficients. The selection of polarization components along single coordinates (by use of spin-precession devices such as Wien filters) and of pure vector or tensor polarizations (by using provisions of the type of source involved) lead to strong simplifications. An example is polarization in the z direction, leaving only the terms $C_{zz,zz}$ and/or $C_{z,z}$ (terms such as $C_{z,zz}$ or $C_{zz,z}$ as well as the analyzing powers $A_z(b)$ and $A_z(t)$ as well as $A_x(b)$ and $A_x(t)$ are forbidden under parity conservation).

The main difficulties with spin-correlation measurements at these low energies are:

- the low cross-sections.
- The use of solid polarized targets can be excluded because it appears impossible to make them sufficiently thin. Therefore only two interacting polarized beams may be employed resulting in low target densities and small yields.
- The use of (compressed) polarized gas at these low energies meets the difficulties of the need for a container including very thin and, at the same time, strong windows of polarization-conserving materials.

Thus, the only sensible experimental arrangement for measuring spin correlations for the D + D reactions is using an intense atomic beam of polarized deuterons as target which is crossed by another atomic or ion beam of polarized deuterons. Alternatively, one could think of building a low-energy storage-ring device in analogy to COSY-Jülich where multiple target passes would compensate for the low cross-sections. However, the technical and financial requirements on such a device seem prohibitive. The high forward multiple-scattering cross-section e.g. requires extremely good vacuum. With existing (decommissioned) polarized-ion sources an experiment can be set up with relatively modest efforts such that acceptable count rates result. Fig. 14.3 sketches such an experimental setup. Besides and after clearing the nuclear-physics questions concerning the low-energy DD reactions many other problems such as preparation of polarized fusion fuel, its injection into magnetic fields and the conservation of polarization have to be investigated.

References

1. Kulsrud, R.M., Furth, H.P., Valeo, E.J., Goldhaber, M.: Phys. Rev. Lett. **49**, 1248 (1982)
2. Kulsrud, R.M.: Nucl. Instrum. Methods Phys. Res. A **271**, 4 (1988)
3. Tanaka, M. (ed.): Proceedings of RCNP Workshop on Spin Polarized Nuclar Fusions (POLU-SION99) RCNP, Osaka (1999).
4. Paetz gen. Schieck, H.: Eur. Phys. J. A **44**, 321 (2010)
5. Honig, A., Sandorfi, A.: In: Imai, K., Murakami, T., Saito, N., Tanida, K., (eds.): Proceeding of the 17th international spin physics symposium (SPIN 2006), Kyoto 2006. AIP Conf. Proc. 915, p. 1010, New York (2007)
6. Didelez, J.-P., Deutsch, C.: EPJ Web of Conferences **3**, 04018 (2010)
7. Deltuva, A., Fonseca, A.: Phys. Rev. C **81**, 054002 (2010)
8. Huke, A., Czerski, K., Heide, P., Ruprecht, G., Targosz, N., Żebrowski, W.: Phys. Rev. C **78**, 015803 (2008)
9. Leemann, Ch., Bürgisser, H., Huber, P., Rohrer, U., Paetz gen. Schieck, H., Seiler, F.: Helv. Phys. Acta **44**, 141 (1971)
10. Leemann, Ch., Bürgisser, H., Huber, P., Rohrer, U., Paetz gen. Schieck, H., Seiler, F.: Ann. Phys. (N.Y.) **66**, 810 (1971)
11. Geist, W.H., Brune, C.R., Karwowski, H.J., Ludwig, E.J., Veal, K.D., Hale, G.M.: Phys. Rev. C **60**, 054003 (1999)
12. Braizinha, D., Brune, C.R., Eiró, A.M., Fisher, B.M., Karwowski, H.J., Leonard, D.S., Ludwig, E.D., Santos, F.D., Thompson, I.J.: Phys. Rev. C **69**, 024608 (2004)

13. Hofmann H.M., Proceedings of Models and Methods in Few-Body Physics, Lisboa, 1986. In: Ferreira, L.S., Fonseca, A.C., Streit, L. (eds.) Lecture Notes in Physics **273**, p. 243, Springer, Berlin (1987)
14. Lemaitre, S.: Diploma thesis, Universität zu Köln, unpublished (1989)
15. Lemaître, S., Paetz gen. Schieck, H.: Few-Body Systems. **9**, 155 (1990)
16. Lemaître, S., Paetz gen. Schieck, H.: Ann. Phys. (Leipzig) **2**, 503 (1993)
17. Geiger, O., Diploma thesis, Universität zu Köln, unpublished (1993)
18. Geiger, O., Lemaître, S., Paetz gen. Schieck, H.: Nucl. Phys. **A586**, 140 (1995)
19. Deltuva, A., Fonseca, A.: Phys. Rev. C **76**, 021001(R) (2007)
20. Hale, G., Doolen, G.: LA-9971-MS. Los Alamos, (1984)
21. Fletcher, K.A., Ayer, Z., Black, T.C., Das, R.K., Karwowski, H.J., Ludwig, E.J., Hale, G.M.: Phys. Rev. C **49**, 2305 (1994)
22. Zhang, J.S., Liu, K.F., Shuy, G.W.: Phys Rev. Lett. **55**, 1649 (1985)
23. Zhang, J.S., Liu, K.F., Shuy, G.W.: Phys. Rev. Lett. **57**, 1410 (1986)
24. Fick, D., Hofmann, H.M.: Phys. Rev. Lett. **55**, 1650 (1983)
25. Hofmann, H.M., Fick, D.: Phys. Rev. Lett. **52**, 2038 (1984)
26. Hofmann, H.M., Fick, D.: Phys. Rev. Lett. **57**, 1410 (1986)
27. Uzu, E., Oryu, S., Tanifuji, M.: In: [3] p. 30 (1999)
28. Uzu, E.: arxiv:nucl-th/0210026 (2002)
29. Zhang, J.S., Liu, K.F., Shuy, G.W.: Phys. Rev. C **60**, 054614 (1999)

Chapter 15
Outlook

The measurement of spin-polarization observables over more than 50 years of nuclear physics, together with the development of ever more sophisticated devices and methods, has proven to deliver essential information especially on reaction mechanisms, often not obtainable by other means. Although a certain saturation in the increase of polarized-beam intensities is visible the quest for higher currents or target densities will continue especially in view of the measurement of more sophisticated experiments such as spin correlations in the astrophysical energy range.

Some measurements of symmetry violations which can only be performed with polarized particles are intended to lower the upper limits of these violations. Examples are direct measurements of quantities sensitive to time-reversal-symmetry violations. One such project is the measurement of electric dipole moments, not only on neutral particles like the neutron, but on charged particles such as the proton or the deuteron. For these experiments high-intensity polarized beams and very sensitive and precise polarization-measurement techniques are required. For these—if done in special storage-ring accelerators—sensitivities better than those of existing measurements by up to two orders of magnitude seem possible, see e.g., Ref. [1]. The basic idea is that the precession of highly spin-polarized particles in the fields of the deflection magnets of a synchrotron, which is exactly known even after very many orbits, would be influenced by electric fields $\vec{E} = \gamma \vec{v} \times \vec{B}$ if the electric dipole moment is $\mu_e \neq 0$ leading to a precession out of the synchrotron plane. The quantity μ_e is forbidden by parity conservation as well as time-reversal invariance but—due to CP non-invariance and in the framework of CPT—could and should be violated even in the standard model and other theories, see e.g., [2]. Finding such a dipole moment $\neq 0$ would have serious consequences for the fundamental theories of the universe. The effects of $\vec{E}$, a change of polarization, can be measured with very high sensitivity and would make such a measurement superior to present determinations of μ_e of the neutron. Details of one such project of the EDM collaboration with an expected sensitivity of 10^{-27} e · cm can be found e.g., in Ref. [3]. In the meantime improvements by two orders of magnitude have been proposed.

H. Paetz gen. Schieck, *Nuclear Physics with Polarized Particles*,
Lecture Notes in Physics 842, DOI: 10.1007/978-3-642-24226-7_15,
© Springer-Verlag Berlin Heidelberg 2012

As pointed out in Ref. [4] normally time-reversal invariance establishes relations between observables of a nuclear reaction in one direction with observables of the inverse reaction, e.g., cross sections ("detailed balance") or outgoing polarizations with analyzing powers which make the relative normalization one of the difficulties of these experiments. Therefore, a "null experiment", i.e., the measurement of *one* quantity alone is the better choice. One such T-odd P-even quantity has been identified, the spin-correlation coefficient $C_{x,yz}$ of elastic scattering of a vector-polarized spin-1/2 beam with a spin-1 tensor-polarized (p_{xz}) target (or vice versa) under $\theta = 0°$, i.e., a total cross-section correlation measurement is necessary. Thus, a polarized proton beam and a tensor-polarized deuteron gas (or jet/storage-cell) target must be used in a transmission experiment. The required sensitivity as well as other experimental conditions will only be met in a storage-ring accelerator such as COSY. An experiment "TRI(C)" was being planned, see e.g., [5].

An entirely new front of research is opening up with the possibility of polarizing antiprotons, e.g., at the facility under construction FAIR at GSI/Darmstadt. The existing results for the nucleon–nucleon interaction will be compared to nucleon-antinucleon data which can shed new light on the nucleon–nucleon and the quark-gluon forces, complemented by annihilation terms. New tests of the standard model and the CPT theorem may emerge. The methods for polarizing antiprotons will be different since Stern-Gerlach devices first require a beam of cold antiatoms, but spin filtering by interaction with polarized beams or target seems viable (but requires maximal beam intensities with high polarization). The method has been confirmed for protons and will soon be implemented at the CERN AD (antiproton decelerator; for a survey see e.g., [6]). It is, however, quite interesting that very recently the formation and trapping of a beam of cold antihydrogen atoms in a cusp trap was successful [7, 8] with the possibility of extracting a beam of highly spin-polarized antiprotons. The main goal is to measure the 1S–2S energy separation as well as studying the hyperfine interaction by doing laser and microwave spectroscopy with similar precision as with H or D atoms. Methods of performing spectroscopic measurements on hydrogen/deuterium using a spinfilter (the central component of Lambshift sources or Lambshift polarimeters, see above) have been proposed recently and seem applicable to antiatoms [9, 10].

References

1. Lebedev, O., Olive, K.A., Pospelov, M., Ritz, A.: Phys. Rev. D **70**, 016003 (2004)
2. Sakharov, A.D.: JETP Lett. **5**, 24 (1967)
3. Aoki M., et al.: dEDM collaboration, AGS Proposal BNL (2008). http://www3.bnl.gov/muon_edm/Deuteron_EDM/EDM_BNL_PRO/deuteron_proposal_080405_1.pdf
4. Conzett, H.E.: Phys. Rev. C **48**, 423 (1993)
5. Bisplinghoff, J., Ernst, J., Eversheim, P.D., Hinterberger, F., Jahn, R., Conzett, H.E., Beyer, M., Paetz gen. Schieck, H., Kretschmer, W.: arxiv:nucl-ex/9810003 (1998)
6. Lenisa, P., Rathmann, F.: CERN Courier **50**(6), 21 (2010)
7. Mohri, A., Yamazaki, Y.: Europhys. Lett. **63**, 207 (2003)
8. Enomoto, Y. et al: (ASACUSA collaboration): Phys. Rev. Lett. **105**, 243401 (2010)

9. Engels, R., Grigoriev, K., Mikirtytchyants, M., Paetz gen. Schieck, H., Rathmann, F., Schug, G., Ströher, H., Vasilyev, A., Westig, M.: Hyperfine Interact. **193**, 341 (2009)
10. Westig, M., Engels, R., Grigoriev, K., Mikirtytchyants, M., Rathmann, F., Paetz gen. Schieck, H., Schug, G., Vasilyev, A., Ströher, H.: Eur. Phys. J. D **57**, 27 (2010)

Index

$\overrightarrow{^3\text{He}}$
  lung imaging, 162
  metastability-exchange
     MEOP, 132
  spin exchange SEOP, 132
^3He
  hyperpolarized, 161
^{129}Xe
  hyperpolarized, 161

A
Analyzing powers, 53–54, 58
  calibration, 145
  coordinate systems, 52
  generalized, 58, 66
  longitudinal, 60

B
Breit–Rabi
  diagram, 109, 117
  formula, 79, 86
  polarimeter, 154

C
Cartesian
  tensors, 29–31, 37, 42, 47, 51
  notation, 38, 51
  basis, 25
  coordinate system, 19, 57
Charge exchange
  H(2S) + Ar, 117
  H(2S) + I$_2$, 117

alkali, 103
  resonant, 104
Conservation
  angular momentum, 57
  isospin, 71
  parity, 57
Coordinate systems, 39, 60
  helicity, 57
  notation, 52
  space-fixed, 54
  transverse, 149
Coulomb barrier, 165
Cross-section
  electron capture in Ar, 120
  electron loss in I$_2$, 117
  electron impact
    ionization, 103, 105
  metastable production
    in Cs, 114
  neutral Cs on H or D, 105

D
Density matrix, 4, 13, 24
  entrance channel, 48
  exit channel, 48
  expansion, 24, 34
    Cartesian or spherical
     tensors, 47
  incident, 24
  rotation, 22, 41
  $S = 1$, 35
  $S = 1/2$, 34
Density operator, 4, 13
Double scattering, 145

H. Paetz gen. Schieck, *Nuclear Physics with Polarized Particles*,
Lecture Notes in Physics 842, DOI: 10.1007/978-3-642-24226-7,
© Springer-Verlag Berlin Heidelberg 2012

E

Ensemble, 9–10
average, 13
statistical, 4
unpolarized, 27
weighted state average, 85
Entropy, 16
Expansion
partial waves, 65

F

Fusion neutrons
DD neutrons, 168
Fusion reactions
five-nucleon reactions, 166
four-nucleon reactions, 167

I

Ionization
ionization yield, 89
Ionizers, 102

M

M or transfer matrix
spin 1/2 + 0, 60
spin 1/2 + 1/2 elastic, 61
Magnetic moments
effective, 86
Magnets
spin-state separation, 77
Stern–Gerlach, 84, 90
matching, 86
Mixed state, 10, 14–15
Multipole fields, 86
Multipole magnets, 11

N

Nuclear g factors, 76, 122, 161

O

Observables, 24
analyzing power, 36, 38, 49–50
Cartesian, 59
one-spin, 50
polarization, 39
polarization transfer, 51, 60
spin correlation, 51
three- and four-spin, 51
two-spin, 51

unpolarized differential, 48
cross-section
zero-spin, 50
Occupation numbers, 20, 83, 100, 155
Operator
expectation value, 9, 24
parity, 58
Pauli, 49
projection, 14
rotation, 39
spin, 18, 20
time-reversal, 57–58
vector, tensor, 30–31
Optical pumping, 131
^{3}He, 131
alkali
Rb, 107, 134

P

Parity
conservation, 23
three-particle reactions, 59
violation, 3, 58
Polarimetry, 145
deuteron polarimeter, 153
deuteron tensor polarimeter, 124, 143
low-energy deuterons, 150, 153
polarimeter
Breit–Rabi, 154
Lambshift, 154
proton polarimeter, 151
Polarization observables, 24
Cartesian, 5
expansions, 4
general, 50
measurement, 143
parity, 23
sensitivities, 3
spherical, 5
symmetry violations, 175
types, 50
Polarization transfer
coefficients, 49, 51, 59
coordinate systems, 54
Wolfenstein
parameters, 51, 61
Polarized ^{3}He
polarized ^{3}He beams, 135
polarized target, 131
Polarized antiprotons
projects, 141, 176
spin filtering, 141
Polarized neutrons

$\overrightarrow{^3\text{He}}$ spin filter, 131
$\overrightarrow{^3\text{He}}$ spin filtering, 141
polarization transfer reactions, 139
Polarized targets
  gas and jet, 125
    storage cell, 126
Polarized-ion sources
  ABS
    ^{7}Li and ^{23}Na beams , 106
    beam formation, 88
    colliding-beams ionizers, CBS, 104
    dissociator, 88
    ECR ionizers, 104
    electron bombardment ionizers, 102
    mutipole fields, 90
    mutipole magnets of Halbach design, 91
    mutipole magnets of modified Halbach design, 92
    separation magnets, 90
    trajectories, 93
  atomic beam, ABS, 87
  ground-state atomic beam, ABS, 88
  Lambshift, 108
    ionization in argon and iodine, 117, 120
    metastables, 114
    Sona scheme, 109, 115
    spin filter, 109
  optical pumping, 107
  spin rotation
    deflection devices, 123
    rotatable Wien filter, 123
Pure state, 4, 13
  basis, 10
  superposition, 10, 18

Q
Quintet suppression factor QSF, 168

R
RF transitions, 93
  adiabatic fast passage, 94
  Majorana formula, 95
  MFT, 100
  quantum-mechanical treatment, 98
  rotating reference
    classical, 94
    quantum-mechanical, 98
  SFT, 100
  SFT condition, 100
  weak-, medium- and strong-field transitions, 95

WFT, 95
  quantum-mechanical, 99
  WFT conditions, 96
Rotation function, 40
  reduced, 41
Rotational symmetry, 76
Rotations, 39
  finite, 39
  infinitesimal, 39
  spherical tensors, 25

S
Sona, 109, 111
  magnets, 111, 115
  transition, 116
spherical harmonics, 5, 32, 36, 41, 84
spherical tensors, 25, 31–32, 34, 41–42
spin correlations, 49
  $\overrightarrow{^3\text{He}}(\vec{d}, p)^4$He, 167
  coordinate system, 54
  cross-section, 170
  D + D, 170
  definition, 51
Spin precession, 39, 43, 121, 124
  curves, 125
  in beam lines and Wien filters, 121
  Larmor frequency, 121
Spin tensor moments, 24
Spin tensors in entrance and exit channels, 47, 65
State
  Zeeman, 56, 80, 118
Stern–Gerlach, 86
  experiment, 75–76
  filter as polarizer, 77
  analyzer, 10
  magnet, 11
  quadrupole, 86
  sextupole, 86
Symmetries
  mirror, 57
  rotational, 57
  time reversal, 145
    time-reversal operator, 57
    time-reversal invariance, 59

T
Tensor moments, 34–35
  notation, 52, 66
  nuclear reactions, 43
  parity behavior, 58

T (*cont.*)
   rotation, 41
   $S = 1$, 35
   $S = 1/2$, 34
  transformation
   parity, 58
   properties, 31
   rotation, 9, 24
   tensors, 5
   unitary, 98

W
Wien filter, 43, 123

Z
Zeeman
   fine structure, 76
   hyperfine splitting, 80
   hyperfine splitting of D atom, 82
   hyperfine splitting of H atom, 81
   hyperfine structure, 77
   lifetime of $n = 2$ states, 109